PE Chemical

Practice Exam

Marta Vasquez, PhD, PE and Robert R. Zinn

Professional Publications, Inc. • Belmont, California

Benefit by Registering This Book with PPI

- Get book updates and corrections.
- Hear the latest exam news.
- Obtain exclusive exam tips and strategies.
- Receive special discounts.

Register your book at **ppi2pass.com/register**.

Report Errors and View Corrections for This Book

PPI is grateful to every reader who notifies us of a possible error. Your feedback allows us to improve the quality and accuracy of our products. You can report errata and view corrections at **ppi2pass.com/errata**.

PE CHEMICAL PRACTICE EXAM

Release History

date	edition number	revision number	update
Oct 2017	1	1	New book.

© 2018 Professional Publications, Inc. All rights reserved.

All content is copyrighted by Professional Publications, Inc. (PPI). No part, either text or image, may be used for any purpose other than personal use. Reproduction, modification, storage in a retrieval system or retransmission, in any form or by any means, electronic, mechanical, or otherwise, for reasons other than personal use, without prior written permission from the publisher is strictly prohibited. For written permission, contact PPI at permissions@ppi2pass.com.

Printed in the United States of America.

PPI
1250 Fifth Avenue, Belmont, CA 94002
(650) 593-9119
ppi2pass.com

ISBN: 978-1-59126-539-9

Library of Congress Control Number: 2017954345

F E D C B A

Table of Contents

Preface and Acknowledgments

To become licensed as a professional chemical engineer, you must pass the Principles and Practice of Engineering Examination—more commonly known as the PE Chemical exam—which is administered by the National Council of Examiners for Engineering and Surveying (NCEES). We wrote *PE Chemical Practice Exam* to be an accurate representation of this important exam. Though you won't find any exact exam problems in this book, its problems closely reflect the content, format, level of difficulty, and length of the problems you will encounter on the morning and afternoon sessions of the PE Chemical exam. In creating such problems, we hope to have provided you with a realistic simulation of the actual exam.

The staff at PPI was essential in guiding this book from its first draft to the completed edition you are now reading. In particular, our thanks go to the PPI Acquisitions and Publishing Services teams, including Steve Buehler, director of acquisitions; Nicole Evans, senior acquisitions editor; Grace Wong, director of publishing services; Cathy Schrott, production services manager; Rebecca Morgan, editorial project manager; Tom Bergstrom, technical illustrator; Richard Iriye and Robert Genevro, typesetters; and Jenny Lindeburg King, editor-in-chief.

Thanks as well to John R. Richards, PhD, PE, and Larry Rotter, PE, who technically reviewed this manuscript, and to Alex Joseph Bertuccio, EIT, Kodi Jean Church, PE, L. Adam Williamson, who double-checked a number of the problems for accuracy, and to Ralph Arcena, calculation checker, for their support in producing this book.

Many people helped with this book, but any errors you may find are ours alone. If you think you have found an error, we hope you will help us keep improving this book by letting us know about it. Please use the errata section of PPI's website, at **www.ppi2pass.com/errata**, to let us know. Valid errata submitted to us will be posted on the website and corrected in future printings and editions of this book.

Marta Vasquez, PhD, PE
Robert R. Zinn

Introduction

ABOUT THIS BOOK

PE Chemical Practice Exam matches the format and specifications of the PE Chemical exam, which is administered by the National Council of Examiners for Engineering and Surveying (NCEES). Like the PE Chemical exam, this book's practice exam consists of a morning and an afternoon session. Each session includes 40 multiple-choice problems that match the level of difficulty of the actual exam and that are solvable in an average of six minutes.

Answer keys are provided so you can quickly check to see if your selected answer option is correct. Step-by-step solutions for all problems are also included to provide more comprehensive solving methods.

The only reference allowed to examinees is the *NCEES PE Chemical Reference Handbook*, which contains all of the equations, tables, and figures that will be covered on the exam. The *NCEES Handbook* is extensively referenced in the solutions in these practice exams to help you get familiar with its contents and organization, so that you can more easily find necessary information during the actual exam.

The nomenclature section in this book lists the variables, symbols, and subscripts used in this book. Like the PE Chemical exam, this book uses both customary U.S. units and SI units. The system of units chosen for a specific problem is based on common practice. For example, in actual chemical engineering practice in the United States, SI units dominate certain fields such as chemical reaction engineering, chemical concentrations, and quantitative analysis. Accordingly, this book uses SI units for problems relating to those fields.

In order to provide the closest approximation of what you will see when you take the actual PE Chemical exam, in this practice exam we have followed the *NCEES Handbook* in as many ways as possible. This means that some common units, variables, and subscripts differ from how they are presented in *PE Chemical Review* and its other companion products such as *PE Chemical Practice.*

For example, *PE Chemical Review* consistently uses the variable v for velocity to avoid confusion with volume and specific volume, while the *NCEES Handbook* represents velocity with u in some places, v in others, and V in still others. Similarly, *PE Chemical Review* consistently uses p for pressure to distinguish it from power, while the *NCEES Handbook* uses P for both pressure and power. The pound-mole is always abbreviated lbmol in *PE Chemical Review*, while the *NCEES Handbook* uses lb mole.

In this practice exam, we have followed *NCEES Handbook* usage in cases like these, even when that usage is inconsistent from one part of the *Handbook* to another. This reflects what you are likely to see on the actual exam. If you have prepared yourself well, small inconsistencies in usage such as these will cause you very little trouble on exam day.

ABOUT THE CHEMICAL PE EXAM

The NCEES Principles and Practice of Engineering (PE) exam in chemical engineering consists of two sessions, one taken in the morning and one taken in the afternoon. Each session is four hours long and contains 40 multiple-choice problems. Problems are independent, and each problem statement contains all the information needed to determine the correct option. Problem types include traditional multiple-choice problems, as well as alternative item types (AITs). AITs include, but are not limited to

- multiple correct, which allows you to select multiple answers
- point and click, which requires you to click on a part of a graphic to answer
- drag and drop, with requires you to click on and drag items to match, sort, rank, or label
- fill in the blank, which provides a space for you to enter a response to the problem

NCEES publishes exam specifications, which list the general subject areas covered by the exam, as well as the approximate number of problems on the exam that will be drawn from each area. The PE Chemical exam specifications are as follows.

I. Mass/Energy Balances and Thermodynamics (16–24)

A. Mass Balances (7–10)

- material balances with no reaction (e.g., phase behavior, mass, volume, density, composition, purge, bypass)

- material balances with reaction (e.g., multiple reactions, incomplete reactions, excess reactant, purge, bypass, recycle, combustion)

B. Energy Balances and Thermodynamics (9–14)

- energy balances on non-reactive systems (e.g., sensible heat, latent heat, heat of solution)
- energy balances on reactive systems (e.g., heat of reaction and combinations with sensible heat; latent heat; heat of solution)
- power cycles (e.g., refrigeration, engines, turbines, heat recovery)

II. Heat Transfer (11–16)

A. Mechanisms (6–9)

- heat transfer without phase change (e.g., thermal conductivity, heat capacity, conduction, convection, free/forced heat transfer coefficients/correlations, radiation, combinations thereof)
- heat transfer with phase change (e.g., vaporization and evaporation, condensation, sublimation, crystallization, latent heat)

B. Applications (5–7)

- heat exchange equipment design (e.g., overall heat transfer coefficient, fouling factors, LMTD, F-factor, equipment selection, insulation)
- heat exchange equipment analysis (e.g., pressure drop, fouling effects, performance evaluation (NTU), changes in parameters)

III. Kinetics (8–12)

A. Reaction Parameters (4–6)

- rate equation (e.g., rate constant, activation energy, order of reaction, mechanisms, catalysis)
- chemical equilibria (e.g., temperature and pressure dependence, composition)

B. Reactors (4–6)

- conversion in single reactors (e.g., batch reactors, continuous stirred tank reactors (CSTR), plug flow reactors (PFR))
- conversion in complex reactors (e.g., reactors in series: CSTR and/or PFR, multiphase reactors, fluidized beds, packed beds, recycle, bioreactors)
- yield and selectivity

IV. Fluids (11–16)

A. Mechanical-Energy Balance (8–12)

- flow behavior (e.g., viscosity, velocity, Reynolds number, friction factors, pressure drop in pipes, valves, and fittings, expansion and contraction, porous media, particle dynamics, fluidization, sonic velocity, laminar and turbulent flow, two-phase flow)
- flow applications (e.g., potential and kinetic energy, friction, flow networks, mixing, pumps, NPSH, turbines, compressors, drivers, solids handling)

B. Flow and Pressure Measurement Techniques (3–4)

- flow measurement application (e.g., mass and volumetric meters) and pressure measurement application (e.g., permanent pressure drop, differential pressure devices)

V. Mass Transfer (10–15)

A. Phase Equilibria (4–6)

- ideal systems (e.g., Henry's Law, Raoult's Law, Dalton's Law, ideal gas law, vapor pressure)
- nonideal systems (e.g., activity coefficients, fugacity coefficients, azeotropes, immiscible and partially miscible phases, equations of state)
- phase equilibrium applications (e.g., bubble point, dew point, flash, critical states)

B. Continuous Vapor-Liquid Contactors (5–7)

- material and energy balances for trayed units and packed units (e.g., absorption, stripping, distillation)
- design parameters for trayed units (e.g., minimum flow rates and reflux, minimum and theoretical stages, feed location, tray selection, capacity, efficiency, flooding, dumping, tray hydraulics)
- design parameters for non-trayed units (e.g., minimum flow rates and reflux, minimum stages, theoretical stages, NTU, feed location, packing selection, capacity, efficiency, flooding, pressure drop, mass transfer coefficients, height of transfer units)

C. Miscellaneous Mass Transfer Processes (1–2)

- continuous, batch, and semicontinuous (e.g., drying, membranes, extraction, crystallization, filtration, leaching, humidification, diffusion, adsorption, absorption, stripping, distillation)

VI. Plant Design and Operation (14–21)

A. Economic Considerations (1–2)

- cost estimation and project evaluation (e.g., capital costs, depreciation, operating costs, risk evaluation, optimization, return on investment)

B. Design (7–11)

- process design (e.g., process flow sheets, P&ID, specifications, procedures, modeling, simulation, scale-up, process and product development, boundary conditions)
- process equipment design (e.g., equipment selection, optimization, design temperature, design pressure)
- siting considerations (e.g., security, ingress, egress, plant layout, utilities, natural disasters, human factors)
- instrumentation and process control (e.g., sensors, controller actions, feedback and feed-forward actions)
- materials of construction (e.g., material properties and selection, corrosion considerations)

C. Operation (3–4)

- process and equipment reliability (e.g., testing, preventive maintenance, startup and shutdown procedures, robustness)
- process improvement and troubleshooting (e.g., debottlenecking, experimental design and evaluation, optimization)

D. Safety, Health, and Environment (3–4)

- protection systems (e.g., pressure or vacuum relief valves (safety valves), flares, rupture disks, vents, vacuum breakers, inerting, seal legs, discharge location, configuration, fire protection)
- industrial hygiene (e.g., MSDS, exposure limits and control, noise control, ventilation, personal protective equipment)
- hazard identification and management (e.g., flammability, explosive limits, auto-ignition, reactor stability, process hazard analysis, safety integrity level (SIL), management of change)
- environmental considerations (e.g., emissions evaluation, permitting, pollution prevention, mitigation, waste determination)

HOW TO USE THIS BOOK

When you are ready to take the practice exam, set a timer for four hours and go through the morning session. For each problem, mark your answers on the answer sheet. After completing the morning session, take a one-hour break to simulate the break you will get on your actual exam day. Then, set your timer again for four hours and take the afternoon session.

When you are done, check your answers against the answer keys. Review the solutions to any problems you answered incorrectly or didn't feel confident about, and compare your problem-solving approaches against those given in the solutions. (However, there are often multiple ways to reach the answer to a problem, and the method given in this book may not be the only valid one possible.)

The morning and afternoon sessions of this practice exam are roughly equivalent in difficulty and scope of topics. (This is generally true of the actual exam as well.) Therefore, another way to use this book is to take one half of this exam early in your studies as a pretest to help you determine what areas need the most review, and then take the other half closer to exam day to retest your preparedness.

For additional exam preparation and practice, go to the PPI Learning Hub (**learn.ppi2pass.com**) to find the *PE Chemical Online Practice Exam,* which mimics the online CBT experience of the PE Chemical exam. You'll also find other PE Chemical prep material, including *PE Chemical Review*, *PE Chemical Practice*, and diagnostic exams at the PPI Learning Hub.

References for the Exam

The PE Chemical exam is not based on or tied to any particular codes, standards, or references. Except for the *NCEES PE Chemical Reference Handbook* (*NCEES Handbook*), NCEES does not provide an official list of references you should be familiar with. During the exam, you will only be able to use the onscreen *NCEES Handbook*. However, the following references are those that you may find helpful when preparing for the exam.

REFERENCES

Crane Engineering Division. *Flow of Fluids Through Valves, Fittings, and Pipe* (Technical Paper No. 410). New York, NY: Crane Co.

Felder, Richard M., and Ronald W. Rousseau. *Elementary Principles of Chemical Processes.* Hoboken, NJ: John Wiley and Sons.

Green, Don W., and Robert H. Perry. *Perry's Chemical Engineers' Handbook.* New York, NY: McGraw-Hill.

McCabe, Warren L., Julian C. Smith, and Peter Harriott. *Unit Operations of Chemical Engineering.* New York, NY: McGraw-Hill.

Smith, J. M., H. C. Van Ness, and M. M. Abbott. *Introduction to Chemical Engineering Thermodynamics.* New York, NY: McGraw-Hill.

Van Wylen, Gordon J., and Richard E. Sonntag. *Fundamentals of Classical Thermodynamics.* Hoboken, NJ: John Wiley and Sons.

Nomenclature

A	annual value	\$	\$
A	area	ft^2	m^2
B	bottoms flow rate	lbm/hr or lb mole/hr	kg/h or kmol/h
BHP, BP	brake horsepower	hp	W
BPE	boiling point elevation	°F	°C
c	specific heat (heat capacity)	Btu/lbm-°F	kJ/kg·°C
C	coefficient	–	–
C	concentration	lbm/gal or lb mole/ft^3	kg/L or kmol/m^3
C	conversion rate	Btu/kW-hr	kJ/kW·h
C	cost	\$	\$
C_p, c_p	molar heat capacity	Btu/lb mole-°F	kJ/kmol·K
C_V	control valve coefficient	–	–
d, D	diameter	ft	m
D	depreciation	\$	\$
D	diffusivity	ft^2/sec	m^2/s
D	distillate flow rate	lbm/hr or lb mole/hr	kg/h or kmol/h
E	activation energy	Btu/lb mole	kJ/kmol
E	energy transfer rate	Btu/hr	kJ/h
EC	cost of electricity	\$/kW-hr	\$/kW·h
f	fraction	–	–
f	friction factor	–	–
F	factor	–	–
F	feed flow rate	lbm/hr or lb mole/hr	kg/h or kmol/h
F	mass of feed	lbm	kg
g	gravitational acceleration, 32.2 (9.81)	ft/sec^2	m/s^2
g_c	gravitational constant, 32.2	ft-lbm/lbf-sec^2	n.a.
G	gas flow rate	lbm/hr or lb mole/hr	kg/h or kmol/h
h	convective heat transfer coefficient	Btu/hr-ft^2-°R	W/m^2·K
h	head or head loss	ft	m
h	height	ft	m
h	specific enthalpy	Btu/lbm	kJ/kg
$\bar{h}$	average convection heat transfer coefficient	Btu/hr-ft^2-°R	W/m^2·K
H	enthalpy	Btu	kJ
H	molar enthalpy	Btu/lb mole	kJ/kmol
HV	heating value	Btu/ft^3	kJ/m^3
i	interest rate	–	–
k	ratio of specific heats	–	–
k	reaction rate constant	1/sec or ft^3/ lb mole-sec	1/s or m^3/ kmol·h
k	thermal conductivity	Btu/hr-ft-°R	W/m·K
K	distribution coefficient	–	–
K	equilibrium ratio	–	–
K	minor loss coefficient	–	–
K	rate constant	–	–
K	resistance coefficient	–	–
LC	landfill generating capacity	kW-hr/yr	kW·h/yr
L	length or equivalent length	ft	m
L	liquid	lb mole	kmol
L	liquid flow rate	lbm/hr or lb mole/hr	kg/h or kmol/h
L	service life	yr	yr
LFG	landfill gas production rate	ft^3/yr	m^3/yr
m	mass	lbm	kg
m	slope	–	–
$\dot{m}$	mass flow rate	lbm/sec	kg/s
M	flow rate from mixing point	lb mole/hr	kmol/h
MW	molecular weight	lbm/lb mole	kg/kmol
n	molar flow rate	lb mole/min	kmol/min
n	number of moles	lb mole	kmol

n	number of periods	–	–
n	reaction order	–	–
N	flow rate from reactor	lb mole/hr	kmol/h
N	number of items	–	–
NPSH	net positive suction head	ft	m
Nu	Nusselt number	–	–
p, P	pressure	lbf/ft^2	Pa
P	percentage	%	%
P	precipitation	ft	m
P	present value	$	$
P	temperature effectiveness	–	–
Pr	Prandtl number	–	–
q	heat loss or heat transfer per unit length, unit area, or unit volume	Btu/hr-ft, Btu/hr-ft^2, or Btu/hr-ft^3	W/m, W/m^2, or W/m^3
q_x	rate of heat loss or heat transfer in x-direction	Btu/hr	W
Q	flow rate	gal/min	L/s
Q	heat	kcal	J
Q, $\dot{Q}$	heat transfer rate	Btu/hr	W
Q	volumetric flow rate	ft^3/hr	m^3/h
r	mole fraction recovered	–	–
r	rate of reaction	lb mole/ft^3-hr	kmol/m^3·h
r	ratio	–	–
R	heat capacity rate ratio	–	–
R	manometer reading	ft	m
R	recycle stream flow rate	lb mole/hr	kmol/h
R	reflux ratio	–	–
R	thermal resistance	ft^2-hr-°F/Btu	m^2·K/W
R	underflow ratio	–	–
R^*	universal gas constant	ft-lbf/ lb mole-°R	J/kmol·K
Ra	Rayleigh number	–	–
Re	Reynolds number	–	–
S	mass of solvent	lbm	kg
S	salvage value	$	$
S	stripping factor	–	–
Sc	Schmidt number	–	–
SG	specific gravity	–	–
Sh	Sherwood number	–	–
t	time	sec	s
T	absolute temperature	°R	K
T	temperature	°F	°C
u	velocity	ft/sec	m/s
U	overall heat transfer coefficient	Btu/hr-ft^2-°F	W/m^2·°C
v	specific volume	ft^3/lbm	m^3/kg
v	velocity	ft/sec	m/s
V	molar specific volume	ft^3/lb mole	m^3/kmol
V	vapor flow rate	lbm/hr or lb mole/hr	kg/h or kmol/h
V	volume	ft^3	m^3
$\dot{V}$	volumetric flow rate	ft^3/sec	m^3/s
w	mass fraction	–	–
w	width	ft	m
W, $\dot{W}$	work	Btu/hr	W
x	liquid mass fraction, liquid mole fraction, or moisture content	–	–
x	quality	–	–
X	fractional conversion	–	–
y	vapor mass fraction or vapor mole fraction	–	–
Y	expansion factor	–	–
z	elevation	ft	m

Symbols

α	order of reaction	–	–
α	thermal diffusivity	ft^2/sec	m^2/s
β	ratio of diameters	–	–
β	thermal expansion coefficient	1/°R	1/K
γ_i	mass concentration	lbm/ft^3	kg/m^3
γ	activity coefficient	–	–
δ	difference in stoichiometric coefficients	–	–
δ	thickness	ft	m
ϵ	emissivity	–	–
ϵ	fractional volume change	–	–
ϵ, ε	specific roughness	ft	m
η	efficiency	–	–
λ	latent heat of vaporization	Btu/lbm	kJ/kg

μ	absolute viscosity	lbm/ft-sec	Pa·s
ν	kinematic viscosity	ft^2/sec	m^2/s
ν	specific volume	ft^3/lbm	m^3/kg
ρ	density	lbm/ft^3	kg/m^3
σ	Stefan-Boltzmann constant, 0.171×10^{-8} (5.67×10^{-8})	Btu/ $hr\text{-}ft^2\text{-}°R^4$	$W/m^2\cdot K^4$
τ	space-time	sec	s
ϕ	relative humidity or ratio of pressures	–	–
ω	specific humidity	lbm/lbm	kg/kg

Subscripts

0	initial
a	activation
a	actual
aux	auxiliary
avail	available
avg	average
b	boiling
bar	barometric
B	bottoms
c	combustion, contraction, correction, or crystals
cond	condensation
conv	convection
C	condenser
d	destroyed or discharge
D	distillate
e	ethylene oxide
eq	equivalent
evap	evaporation
exch	heat exchanger
f	firebrick, fixed, formation, or friction
F	feed
g	glycol
G	gas
h	hydraulic or hydrocarbon
horiz	horizontal
i, i	ideal, inflow, inner, or insulation brick
lm	logarithmic mean
L	liquid
man	manometer
max	maximum
min	minimum
noncond	noncondensible
N	stage N
o, o	outer, outflow, or overflow
p	constant pressure, pressure, or pump
precip	precipitation
r	refractory brick
rad	radiation
ref	reference
rem	remaining
rxn	reaction
R	radiation or reboiler
s	settling, sludge, solution, static, steam, styrene, or surface
sat	saturated
std	standard
t	toluene
uncond	uncondensed
v	valve or vapor
vert	vertical
V	vapor stream
w	water
wv	water vapor
z	elevation

Morning Session Answer Sheet

1. (A) (B) (C) (D)	11. (A) (B) (C) (D)	21. (A) (B) (C) (D)	31. (A) (B) (C) (D)
2. (A) (B) (C) (D)	12. ________________	22. (A) (B) (C) (D)	32. (A) (B) (C) (D)
3. (A) (B) (C) (D)	13. (A) (B) (C) (D)	23. (A) (B) (C) (D)	33. drag and drop
4. (A) (B) (C) (D)	14. ________________	24. (A) (B) (C) (D)	34. (A) (B) (C) (D)
5. (A) (B) (C) (D)	15. (A) (B) (C) (D)	25. (A) (B) (C) (D)	35. (A) (B) (C) (D)
6. (A) (B) (C) (D)	16. (A) (B) (C) (D)	26. (A) (B) (C) (D)	36. (A) (B) (C) (D)
7. (A) (B) (C) (D)	17. (A) (B) (C) (D)	27. point and click	37. (A) (B) (C) (D)
8. (A) (B) (C) (D)	18. point and click	28. (A) (B) (C) (D)	38. (A) (B) (C) (D)
9. (A) (B) (C) (D)	19. (A) (B) (C) (D)	29. (A) (B) (C) (D)	39. (A) (B) (C) (D)
10. (A) (B) (C) (D)	20. (A) (B) (C) (D)	30. (A) (B) (C) (D)	40. (A) (B) (C) (D)

Morning Session

1. An irreversible second-order reaction is being carried out in a system consisting of a plug-flow reactor (PFR) followed by a continuous stirred tank reactor (CSTR). Pure species M is fed into the PFR with an initial concentration of 3.25 $kmol/m^3$. The space-time in each of the reactors is 1.3 min, and the reaction rate constant is 0.987 $m^3/kmol{\cdot}min$.

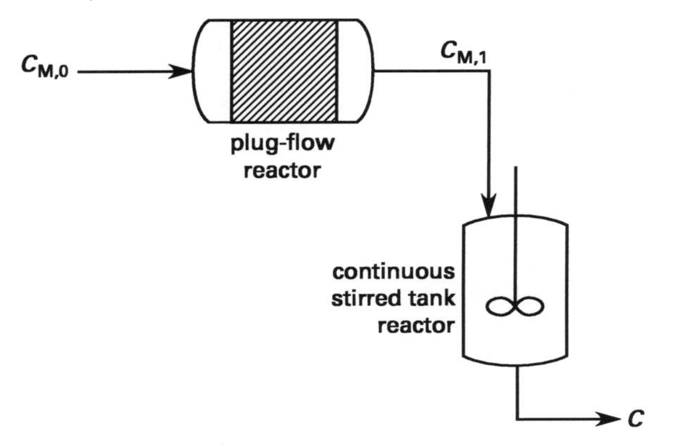

The conversion of species M at the exit of the CSTR is most nearly

(A) 0.40

(B) 0.60

(C) 0.80

(D) 0.90

2. Saturated steam is used in an insulated countercurrent heat exchanger to heat a stream of pure oxygen as shown.

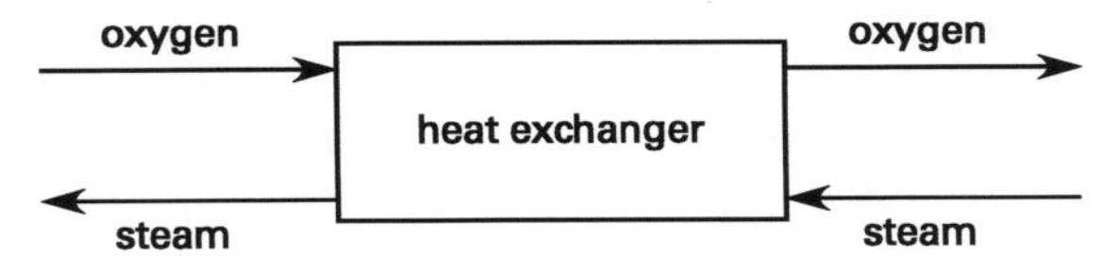

The oxygen flows at a rate of 125 kmol/h and enters the heat exchanger at 30°C. The saturated steam flows at a rate of 225 kg/h. Both the steam and the oxygen are at a constant pressure of 1.4 bar. At these conditions, the latent heat of the saturated steam is 2230.95 kJ/kg. The heat capacity of the oxygen stream is

$$c_p = 29.88\ \frac{\text{kJ}}{\text{kmol}{\cdot}\text{K}} - \left(0.011\,38\ \frac{\text{kJ}}{\text{kmol}{\cdot}\text{K}^2}\right)T$$

The outlet temperature of the oxygen stream is most nearly

(A) 150°C

(B) 190°C

(C) 220°C

(D) 230°C

3. A thermal incinerator consists of a combustion chamber and a waste gas preheater. The incoming waste feed stream, which consists of waste gas and combustion air, passes through the preheater before entering the combustion chamber. While in the preheater, the waste feed is heated by flue gas leaving the combustion chamber. The waste feed enters the preheater at 110°F and leaves it at 1227.5°F. The combustion chamber operates at 1600°F, and the flue gas temperature is reduced to 482.5°F before it leaves the preheater.

The waste feed volumetric rate is 25,000 ft^3/min measured at 77°F and 14.7 lbf/in^2. The waste feed has a density of 0.0739 lbm/ft^3 and a heat of combustion of −58.47 Btu/lbm. The flue gas has the same density as the waste feed. The auxiliary fuel has a density of 0.0408 lbm/ft^3 and a heat of combustion of −19,500 Btu/lbm. The heat capacity of air is 0.255 Btu/lbm-°F, and the heat capacities of both waste and flue gases on both sides of the preheater are approximately the same as that of the air. Energy losses are 12% of the total energy input to the incinerator.

The volumetric flow rate of the auxiliary fuel is small relative to that of the waste gas. Combustion in the preheater is negligible; the mass flow rate for the waste gas is approximately the same both entering and exiting the preheater, and the same is true for the flue gas.

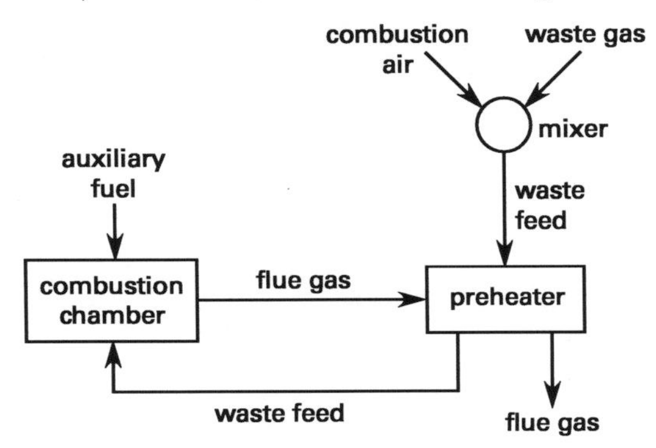

The flow rate of auxiliary fuel measured at 77°F and 14.7 lbf/in^2 is most nearly

(A) 190 ft^3/min

(B) 240 ft^3/min

(C) 300 ft^3/min

(D) 470 ft^3/min

4. A flue gas with a density of 1350 kg/m^3 flows through a steel pipe with an inside diameter of 150 mm. A venturi meter with a throat diameter of 75 mm is installed in the line as shown.

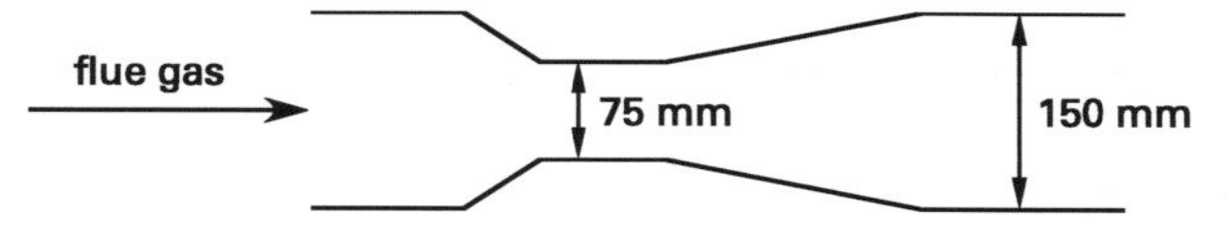

The venturi meter shows a differential column height of 32 mm on a mercury manometer. The density of the water is 1000 kg/m^3, the specific gravity of mercury is 13.6, and the flow coefficient is 0.985. The mass flow rate of the flue gas is most nearly

(A) 0.011 kg/s

(B) 14 kg/s

(C) 20 kg/s

(D) 460 kg/s

5. The two gas phase reactions $M \rightarrow N$ and $Q \rightarrow S + U$ are carried out simultaneously in a plug-flow tubular reactor. The reactor is operated isothermally at 450°F and 5.2 atm. The reaction $M \rightarrow N$ is first-order with a reaction rate constant of 1200 min^{-1}. The reaction $Q \rightarrow S + U$ is zero-order with a reaction rate constant of 3.6 lb mole/ft^3-min. The feed is equimolecular in M and Q and enters at a flow rate of 15 lb mole/sec. Assume a negligible pressure drop and that the ideal gas law applies.

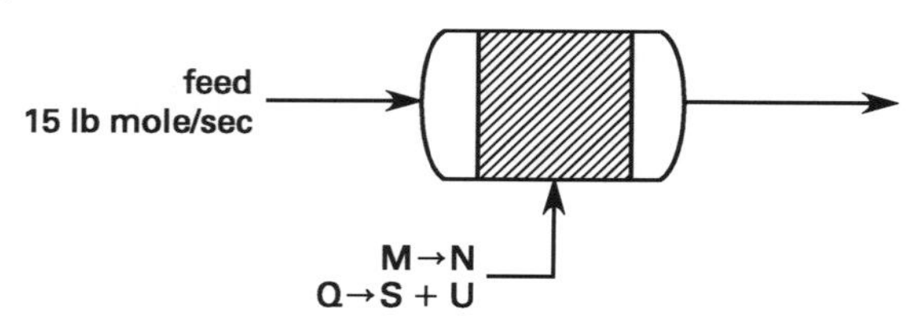

The volume of the plug-flow reactor required to achieve a 60% conversion of M is most nearly

(A) 0.81 ft^3

(B) 100 ft^3

(C) 120 ft^3

(D) 210 ft^3

6. Ambient air enters an adiabatic saturator at 14.7 lbf/in^2 and 75°F, and leaves at 50°F as shown.

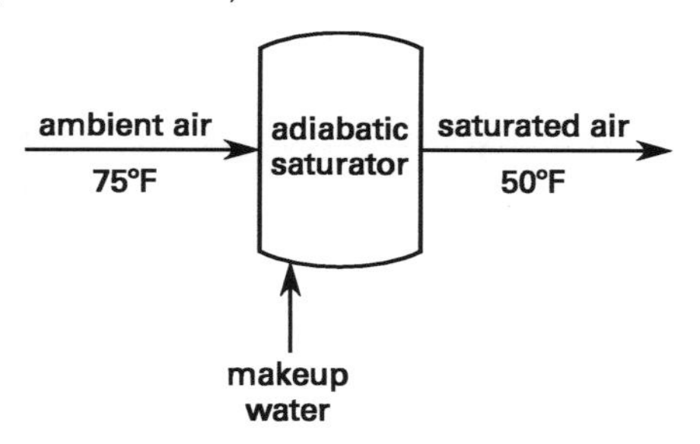

The average heat capacity of dry air is constant at 0.24 Btu/lbm-°F. Assume that the ideal gas law and Dalton's law apply. The relative humidity of the ambient air entering the saturator is most nearly

(A) 0.41

(B) 0.71

(C) 1.4

(D) 33

7. A stream of waste gas at 270°F is fed through a 5 ft diameter smokestack as shown.

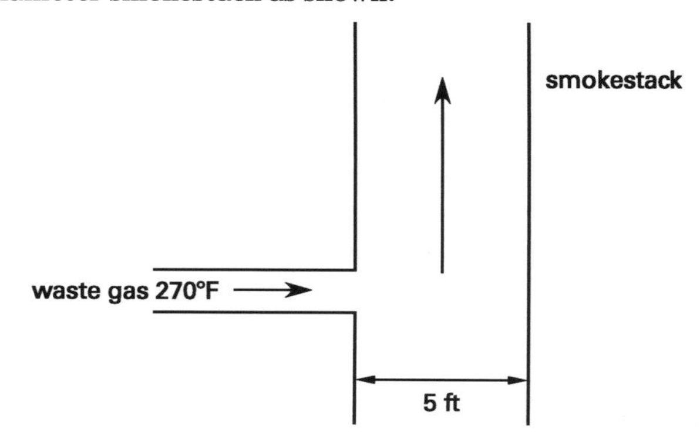

The waste gas has an average molecular weight of 29 lbm/lb mole and contains 0.6% (by weight) sulfur dioxide (SO_2). A pitot tube at the center of the duct registers 0.70 in wg, and a manometer at the duct wall registers −0.60 in wg. Barometric pressure is 725 mm Hg. The molecular weight of the sulfur dioxide is 64 lbm/lb mole. The density of water is 62.4 lbm/ft^3. The average velocity of the waste gas stream is 81% of its maximum velocity. Assume the ideal gas law applies. The mass flow rate of SO_2 in the smokestack is most nearly

(A) 100 lbm/hr

(B) 2600 lbm/hr

(C) 3600 lbm/hr

(D) 4400 lbm/hr

8. A gas stream consisting of nitrogen (N_2) and acetylene (C_2H_2) reacts to produce hydrogen cyanide (HCN) in an adiabatic reactor. An excess of N_2 is used to convert all the C_2H_2.

The reactor outlet stream is separated. The separator produces two streams, one consisting of pure N_2 and the other of pure HCN. The N_2 stream from the separator is recycled and mixed with fresh feed before entering the reactor. The desired production of HCN is 1000 lbm/hr. The maximum capacity of the recycled stream is 52 lbm/hr. The molecular masses are

$$MW_{N_2} = 28 \text{ lbm/lb mole}$$
$$MW_{C_2H_2} = 26 \text{ lbm/lb mole}$$
$$MW_{HCN} = 27 \text{ lbm/lb mole}$$

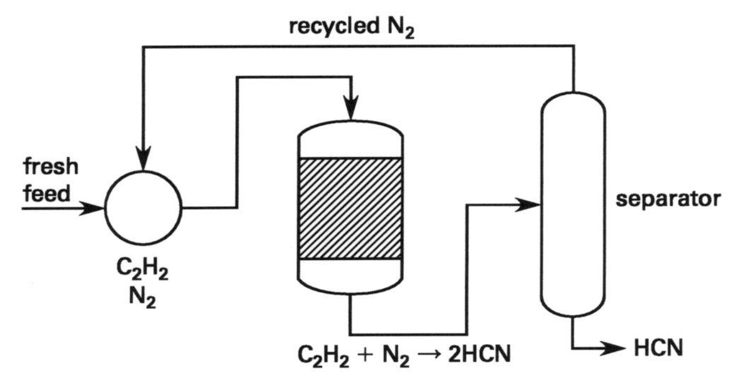

The percentage of excess N_2 in the feed stream necessary to meet the desired production is most nearly

(A) 4.8%

(B) 9.1%

(C) 10%

(D) 110%

9. A 1000 lbm feed slurry consisting of 7.9% (by weight) inert solids, 12% ferric chloride ($FeCl_3$), 7.9% hydrochloric acid (HCl), and the balance water is extracted with pure isopropyl ether in a single pass to recover the $FeCl_3$. 8.9% of the raffinate consists of inert solids. The extract contains no solids. The desired recovery of $FeCl_3$ is 99%. Isopropyl ether and water are immiscible.

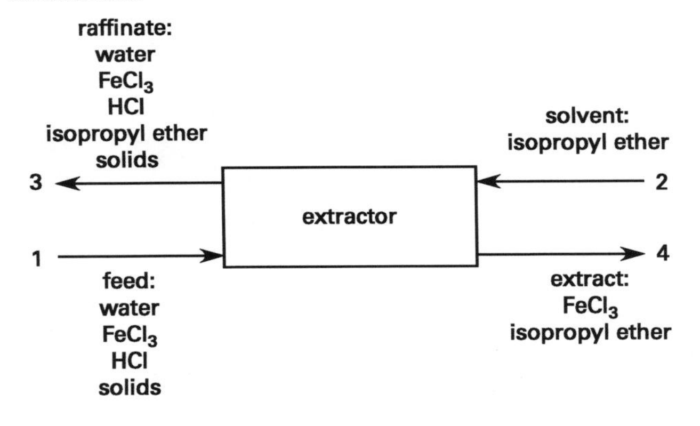

The mass of isopropyl ether required to achieve the desired recovery is most nearly

(A) 29 lbm

(B) 120 lbm

(C) 640 lbm

(D) 760 lbm

10. Formaldehyde (CH_2O) is produced by the partial oxidation of methanol (CH_3OH) in a reactor. The feed stream into the reactor contains 473 lb mole/day of CH_3OH and 899 lb mole/day of air. The conversion of this feed is 60%.

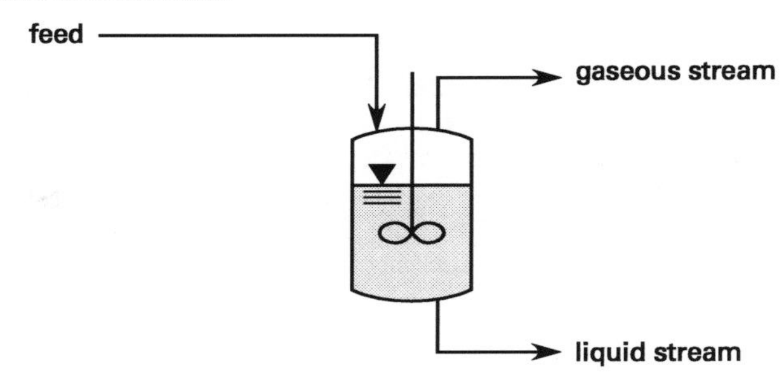

The reactor product is separated into two streams, one gaseous and one liquid. The gaseous stream consists of

- carbon monoxide (CO)
- carbon dioxide (CO_2)
- 5% (by mole) hydrogen (H_2)
- nitrogen (N_2)

The liquid stream consists of

- equal amounts of CH_3OH and CH_2O
- 0.4% (by mole) formic acid (HCOOH)
- water (H_2O)

Assume that the air is 21% (by mole) oxygen (O_2) and 79% (by mole) N_2. The molar flow rate of the water in the liquid stream is most nearly

(A) 51 lb mole/day

(B) 330 lb mole/day

(C) 720 lb mole/day

(D) 840 lb mole/day

11. A furnace wall is 0.48 m high by 3.2 m wide. The wall is constructed from 0.20 m thick firebrick with a thermal conductivity of 2.20 W/m·K. The temperatures of the inner and outer surfaces are 1450K and 1024K, respectively. At steady-state operation, the rate of heat loss through the wall is most nearly

(A) 2200 W

(B) 4700 W

(C) 7200 W

(D) 15 000 W

12. According to the American National Standards Institute (ANSI) International Society of Automation (ISA) standard S5.1-1984 (R 1992) a **primary element** is a synonym for ____________.

Fill in the blank.

13. A binary mixture consisting of pentane and hexane is stored at its bubble point of 151.9°C and at a pressure of 1200 mm Hg. At this temperature, the vapor pressure of pure pentane is 1647.7 mm Hg and that of pure hexane is 785.9 mm Hg. Assume Dalton's law and Raoult's law apply. The mole fraction of pentane in the mixture is most nearly

(A) 0.48

(B) 0.50

(C) 0.52

(D) 0.66

14. According to the American National Standards Institute (ANSI) International Society of Automation (ISA) standard S5.1-1984 (R 1992) a combination of two or more instruments or control functions arranged so that signals pass from one to another for the purpose of measurement and/or control of a process variable is a __________.

Fill in the blank.

15. The irreversible reaction M $\rightarrow$ products was carried out in a batch reactor in order to observe concentration versus half-life. The following data were obtained.

t (min)	C_M (mol/L)
1.7	784
3	258
5.3	85
9.2	28
16	9.2
28	3

The reaction order is most nearly

(A) 1.0

(B) 1.5

(C) 2.0

(D) 2.5

16. A sieve-tray tower recovers ethanol from a gas mixture at a constant pressure of 110 kPa. Acetic acid is used as the solvent. The inlet gas contains 5% (by mole) ethanol and 95% methane. The entering gas molar flow rate is 100 kmol/h. The process is countercurrent, isothermal, and operating at an actual liquid molar flow rate 1.5 times the minimum. At the conditions of the process, the vapor pressure of ethanol is 10.5 kPa, and the liquid-phase activity coefficient of ethanol is 6. Assume no vaporization of absorbent into the carrier gas or absorption of carrier gas by the liquid. The number of stages required to recover 90% of the ethanol is

(A) 2

(B) 3

(C) 4

(D) 5

17. An insulated pipe with an outside diameter of 100 mm carries saturated steam at 200°C. The emissivity of the pipe surface is 0.85. The temperature of the surrounding air is constant at 20°C. The convection heat transfer coefficient from the surface of the pipe to the air is 16.5 W/m^2·K. Assume steady-state conditions. The rate of heat loss from the surface per unit length of pipe is most nearly

(A) 933 W/m

(B) 1580 W/m

(C) 1630 W/m

(D) 31 600 W/m

18. A schematic of a process flowchart is shown.

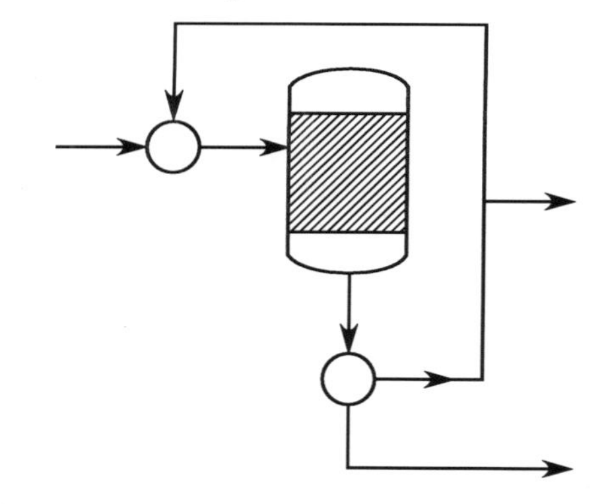

Mark the splitting point on the diagram.

19. A pump supplies water from a large open tank to a scrubbing tower as shown. The water flows through a pipe with an inside diameter of 4.026 in. A control valve keeps the flow rate at 50 gal/min. At this flow rate, the pump head is 136 ft. The pipe has an equivalent length of 248 ft and a roughness of 0.00007 ft. The scrubber operates at atmospheric pressure and its inlet, at point B, is 65 ft above the level of the water in the tank, at point A. The water is at 80°F with a density of 62.2 lbm/ft^3 and a viscosity of 0.86 cP.

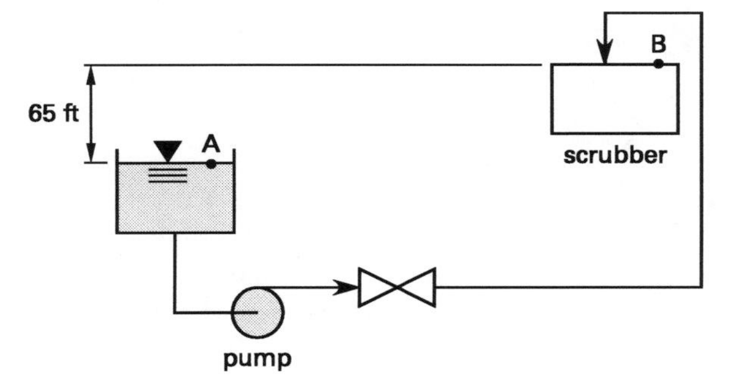

The control valve head is most nearly

(A) −200 ft

(B) −65 ft

(C) 71 ft

(D) 140 ft

20. A conditioned space is to be maintained at 75°F dry-bulb temperature and 50% relative humidity by a refrigeration cooling coil in an adiabatic forced air system. The total cooling load on the conditioned space is 65,000 Btu/hr. Outside air at 96°F dry-bulb and 80°F wet-bulb temperature is provided at a mass rate that is 28% of the supply air rate. Air leaving the cooling coil is at a temperature of 50°F dry-bulb and 86% relative humidity.

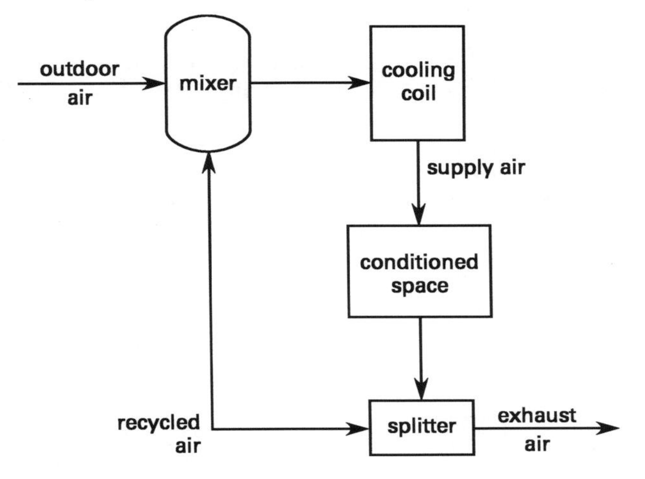

The cooling coil load is most nearly

(A) 5.0 tons of refrigeration/hr

(B) 7.4 tons of refrigeration/hr

(C) 12 tons of refrigeration/hr

(D) 44 tons of refrigeration/hr

21. A distillation column is equipped with a partial reboiler and total condenser. It operates at steady state and is designed to continuously distill 100 kmol/h of a liquid mixture of 49.96% acetone and 50.04% water (by mole). The distillation column also receives another feed stream that consists of 20 kmol/h of pure diglyme. The column produces 50 kmol/h of distillate consisting of 99.90% acetone, 0.05% water, and 0.05% diglyme. The reflux ratio is 3. The operating line equation for the stripping section of the column is most nearly

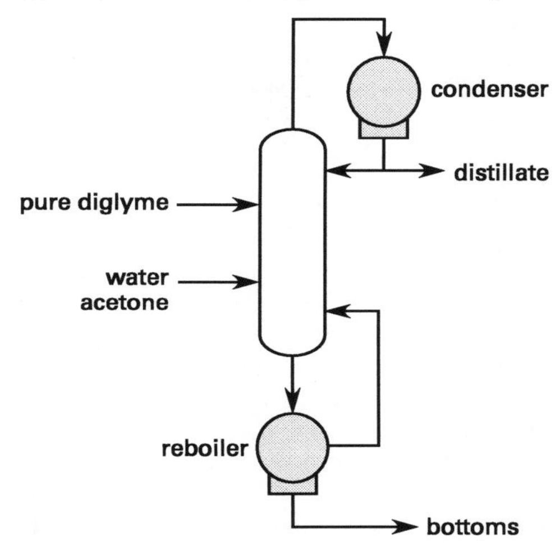

(A) $y = 0.75x + 0.000050$

(B) $y = 0.75x + 0.25$

(C) $y = 1.25x + 0.000050$

(D) $y = 1.35x + 0.000050$

22. Coal with sulfur content of 3.5% (by weight) is burned at a rate of 68 kg/min in a combustion chamber as shown. The sulfur in the ash is 6% (by weight) of the sulfur in the feed. The molecular weight of sulfur is 32 kg/kmol. The molecular weight of sulfur dioxide is 64 kg/kmol.

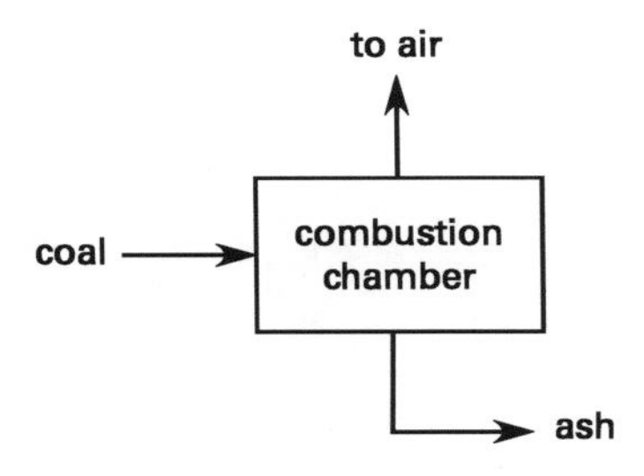

The annual rate of emission of sulfur dioxide to the air is most nearly

(A) 5.9×10^5 kg/yr

(B) 1.3×10^6 kg/yr

(C) 2.4×10^6 kg/yr

(D) 7.5×10^6 kg/yr

23. A municipal waste plant is being designed to produce 3.60×10^6 m^3 of a landfill gas every year for the next 20 years. The landfill gas is composed of 60% methane (by volume) and will be used to produce electricity. Pure methane has a heating value of 33 810 kJ/m^3. The internal combustion engine that will be used requires 14 235 kJ to generate 1 kW·h of electricity. The electricity can be sold for \$0.08/kW·h. Assuming an annual interest rate of 7%, the present value of the investment is most nearly

(A) \$40,000

(B) \$4.0 million

(C) \$7.0 million

(D) \$20 million

24. Two tanks of equal size are arranged in sequence as shown. The tanks are used to preheat a liquid. The liquid is fed into the first tank at 210 kg/min, and it overflows into the second tank and then leaves the second tank at the same flow rate. Each tank is initially filled with 2300 kg of liquid at 25°C, and the liquid fed to the first tank is at 25°C. Saturated steam at a temperature of 320°C condenses in coils immersed in each tank. The coils in each tank have a total heat transfer area of 1 m^2.

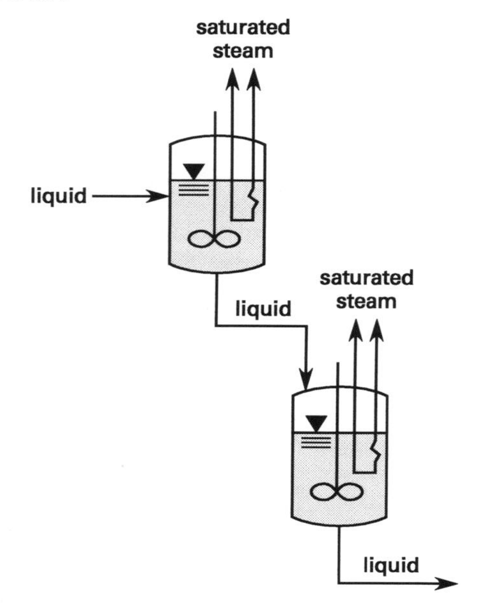

Heat is transferred from the coils to the liquid at a rate that is proportional to the overall heat transfer coefficient, the area of heat transfer, and the difference of temperatures between the steam and the liquid. The liquid has an average constant heat capacity of 2.03 kJ/kg·°C. The overall heat transfer coefficient is 15 kJ/m^2·min·°C. Assume that the density of the liquid remains constant and that each tank remains perfectly mixed throughout the process. The steady-state temperature reached in the second tank is most nearly

(A) 25°C

(B) 35°C

(C) 45°C

(D) 170°C

25. In an adiabatic dryer, a countercurrent of air is used to dry a solid polymer product as shown.

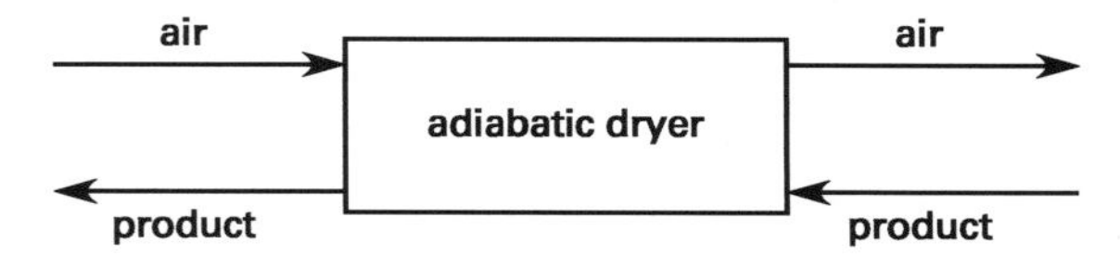

The product enters the dryer at a temperature of 22°C. As it enters, it contains 1.43 g of water for every 1.0 g of dry polymer. The product leaves the dryer at a rate of 345 kg/h at 43°C and containing 7.4% (by weight) water. The heat capacity of the dry polymer is 0.36 J/g·°C, and the heat capacity of the water is 4.184 J/g·°C. The enthalpy change of the product during the drying process is most nearly

(A) 4.0×10^4 J/h

(B) 2.4×10^6 J/h

(C) 3.5×10^7 J/h

(D) 4.0×10^7 J/h

26. Water is heated from 25°C to 68°C by passing it through a thick-walled heated tube. The tube is well insulated, with an inner diameter of 40 mm and an outer diameter of 60 mm. The tube is electrically heated within the wall and provides a uniform heat of 1.5×10^6 W/m^3. The tube's inner surface temperature is 76°C at the outlet. The convection heat transfer coefficient at the outlet is most nearly

(A) 370 W/m^2·K

(B) 440 W/m^2·K

(C) 470 W/m^2·K

(D) 2300 W/m^2·K

27. A schematic of a gas cleaning and particle collection process is shown.

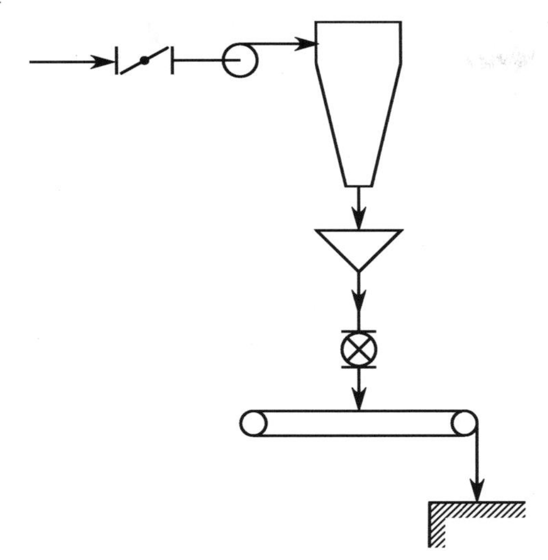

Mark the cyclone separator on the diagram.

28. An open tank contains a liquid with a density of 62.22 lbm/ft^3 and a viscosity of 0.743 cP. This liquid flows through a pipe with an inside diameter of 1.38 in and a specific roughness of 0.00015 ft. When the level of liquid in the tank is 7 ft, the flow rate is 32 gal/min. The system is composed of a standard 90° elbow with an equivalent length to diameter ratio of 30, a half-open gate valve with an equivalent length to diameter ratio of 160, and one globe valve with an equivalent length to diameter ratio of 340. A heat exchanger with an equivalent length of 40 ft is located between the two valves. The length of the horizontal pipe (not including the distances covered by the fittings) is 51 ft. The elevation of the horizontal pipe is 0 ft.

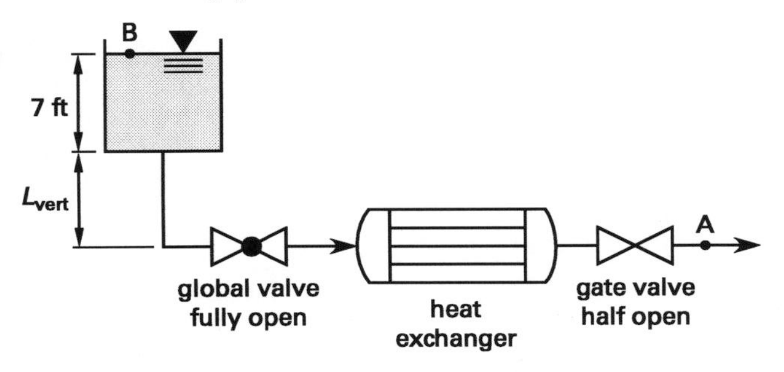

The elevation of the surface of the liquid is most nearly

(A) 8.5 ft

(B) 17 ft

(C) 19 ft

(D) 26 ft

29. Water is pumped through a pipe at a flow rate of 85 gal/min. The water has a density of 62.2 lbm/ft^3 and a viscosity of 0.86 cP, and the pump efficiency is 0.75. The pipe's inside diameter is 2.067 in, and its specific roughness is 0.00018 ft. The pipeline includes four specific 90° threaded elbows with resistance coefficients of 0.57 each (one is on the pump outflow and is not shown), one globe lift check valve with a resistance coefficient of 28, and one gate valve with a resistance coefficient of 0.15.

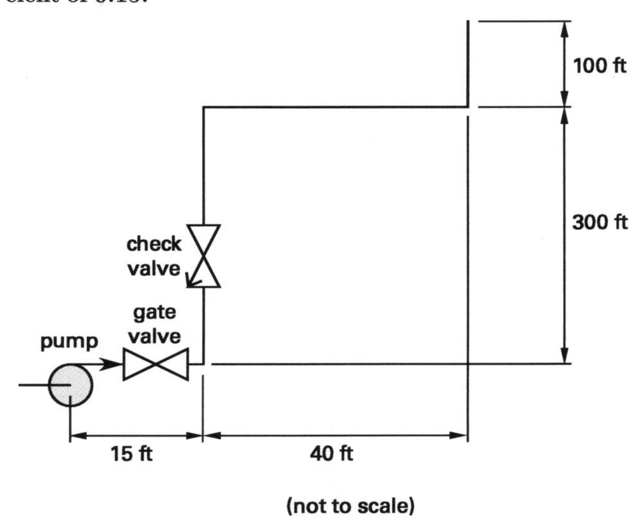

As shown, the pipe system contains two horizontal and two vertical sections. The horizontal sections have equivalent lengths of 15 ft and 40 ft. The vertical sections have equivalent lengths of 300 ft and 100 ft. The resistance coefficient of a sudden exit is 1.0. The total brake horsepower required for the pump is most nearly

(A) 2.6 hp

(B) 10 hp

(C) 11 hp

(D) 14 hp

30. A wastewater stream has a density of 62.4 lbm/ft^3 and a flow rate of 550 gal/min. The stream contains the volatile organic compounds (VOCs) benzene (C_6H_6), toluene (C_7H_8), and ethylbenzene (C_8H_{10}). The VOCs are to be stripped from the wastewater using 3700 ft^3/min of air at 32°F and 14.7 lbf/in^2. The stripping column is operated at 70°F and 15 lbf/in^2, and the number of theoretical stages is 3. The molecular weight of the wastewater is 18.02 lbm/lb mole. The absorption of air by the water and the stripping of water by the air are negligible. The concentrations and the distribution coefficients for the VOCs at 70°F and 15 lbf/in^2 are

component	concentration, C (mg/L)	distribution coefficient, K
C_6H_6	160	256
C_7H_8	45	248
C_8H_{10}	20	285

After the stripping operation, the concentration of the total VOCs remaining in the wastewater is most nearly

(A) 0.0 mg/L

(B) 0.21 mg/L

(C) 2.1 mg/L

(D) 220 mg/L

31. A hydrocarbon lubricant flows in the outer pipe of a double-pipe heat exchanger. The lubricant is cooled from 150°C to 100°C by water flowing through the inner pipe. The water enters the heat exchanger at 60°C and leaves at 80°C. The annular flow is laminar. The outside diameter of the inner pipe, D_{outer}, the inside diameter of the outer pipe, D_{inner}, and the physical properties of the lubricant and water are

characteristic	water	lubricant
diameter of pipe	$D_{\text{inner}} = 0.025$ m	$D_{\text{outer}} = 0.045$ m
mass flow rate	$\dot{m}_w = 0.2$ kg/s	$\dot{m}_h = 0.1$ kg/s
heat capacity	$c_{p,w} = 4178$ J/kg·°C	$c_{p,h} = 3342$ J/kg·°C
viscosity	$\mu_w = 725 \times 10^{-6}$ N·s/m^2	$\mu_h = 3.25 \times 10^{-2}$ N·s/m^2
thermal conductivity	$k_w = 0.625$ W/m·K	$k_h = 0.138$ W/m·K
Prandtl number	$Pr_w = 4.85$	$Pr_h = 27\,500$
average Nusselt number	$\text{Nu}_w = ?$	$\text{Nu}_h = 5.56$

The average Nusselt number for the flow of water has not been determined. The thermal resistance of the pipe walls, heat loss to the surroundings, and fouling factors are all negligible, as are the kinetic and potential energy changes. Assume constant physical properties, and fully developed conditions for the water and lubricant.

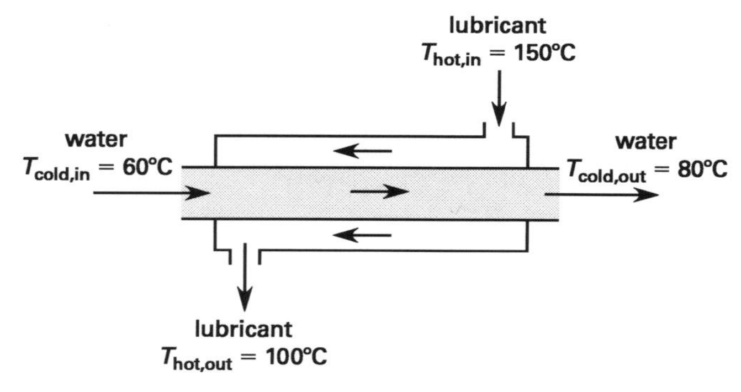

The heat exchanger pipe length required to cool the lubricant is most nearly

(A) 53 m

(B) 58 m

(C) 110 m

(D) 230 m

32. An ideal upward-flow sedimentation tank is shown.

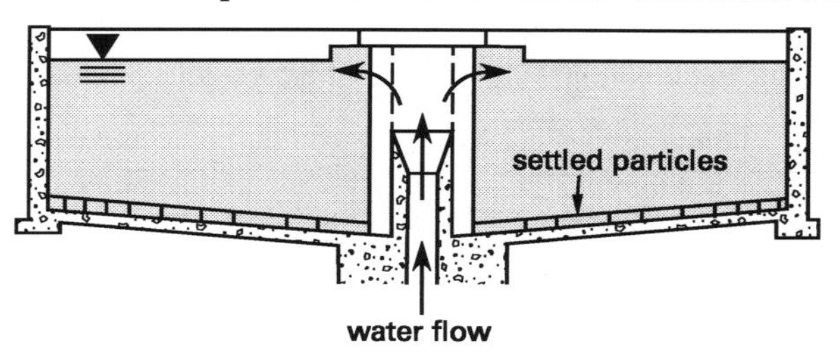

Water flows into the tank at a rate of 15 m^3 per square meter of the tank's surface area. The water contains particles with average settling velocities of 0.1 mm/s, 0.2 mm/s, and 1 mm/s. The percentage of particles with a settling velocity of 0.1 mm/s removed is most nearly

(A) 0.00067%

(B) 0.58%

(C) 60%

(D) 250%

33. A list of symbols and labels are shown using symbology from the American National Standards Institute (ANSI) International Society of Automation (ISA) standard S5.1-1984 (R 1992) and common industry usage.

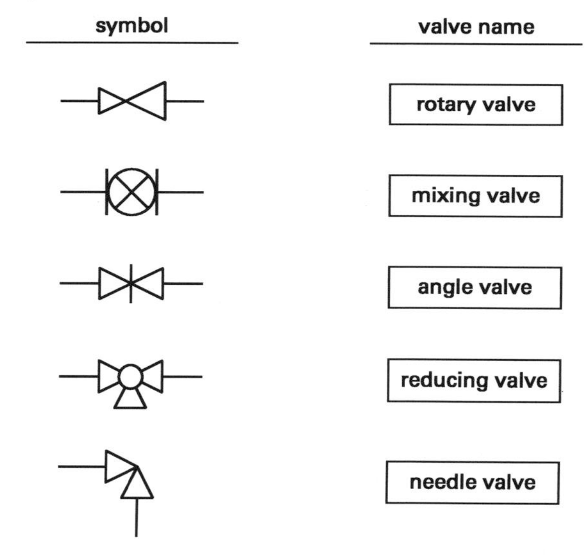

Match the control valve body symbol descriptions to their corresponding names.

34. A solution of starch is concentrated from 6% to 15% (by weight) in a vertical tube evaporator. The solution's boiling point elevation is negligible. Saturated steam at 256°F is used. The temperature corresponding to the pressure in the vapor space is 220°F. The heating surface is 150 ft^2, and the overall heat transfer coefficient is 785 Btu/hr-ft^2-°F. The feed is at a temperature of 220°F. Assume steady-state operation.

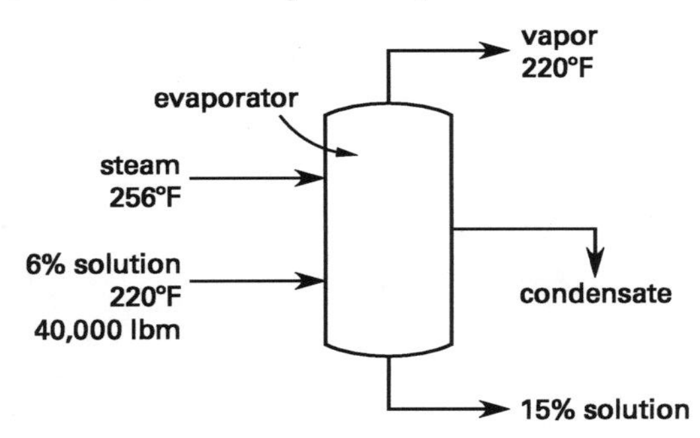

The time it will take to concentrate 40,000 lbm of starch solution is most nearly

(A) 3.1 hr

(B) 3.7 hr

(C) 6.5 hr

(D) 8.6 hr

35. At low temperatures, the rate law for the exothermic irreversible reaction $2M + N \rightarrow 2R$ is

$$-r_M = k_M C_M C_N^{0.5}$$

At high temperatures, the reaction is reversible. The rate law for the reversible reaction is

(A) $-r_M = k_M\left(C_M C_N^{0.5} - \frac{C_R}{\sqrt{K}}\right)$

(B) $-r_M = k_M\left(C_M C_N^{0.5} - \frac{C_R}{K}\right)$

(C) $-r_M = k_M\left(C_M C_N - \frac{C_R^2}{\sqrt{K}}\right)$

(D) $-r_M = k_M\left(C_M^2 C_N - \frac{C_R}{K}\right)$

36. A feed of 267 lb mole/hr, consisting of methanol and water, is to be continuously rectified to produce a distillate at a molar flow rate of 85.1 lb mole/hr and a bottoms stream at a rate of 181.9 lb mole/hr. The feed is to be preheated so that its molar enthalpy is 1189 Btu/lb mole. The molar enthalpy of the saturated vapor is 17,546 Btu/lb mole, that of the saturated liquid is 1782 Btu/lb mole, and that of the bottoms stream is 2580 Btu/lb mole. A reflux ratio of 1.038 will be used. The heat load in the reboiler is most nearly

(A) -3.0×10^6 Btu/hr

(B) 1.4×10^6 Btu/hr

(C) 2.9×10^6 Btu/hr

(D) 3.0×10^6 Btu/hr

37. A pipe with an inside diameter of 2.469 in carries water with a density of 62.4 lbm/ft^3. A pitot tube and a mercury manometer are installed to measure the mass flow rate of the water through the pipe. For the existing steady-state conditions, the manometer reads 1.65 in. The specific gravity of mercury is 13.6. Assume that the average velocity of the water is 96.9% of the velocity in the center of the pipe. The mass flow rate of the water is most nearly

(A) 1300 lbm/min

(B) 4600 lbm/min

(C) 9800 lbm/min

(D) 15,000 lbm/min

38. The single-effect evaporator shown is used to concentrate 15 000 kg/h of a weak liquor solution from a concentration of 1.5% to 25% (by weight). The feed enters the evaporator at 25°C, and saturated steam enters at 120°C. The pressure in the evaporator is 22.88 kPa, and the overall heat transfer coefficient is 2750 W/m^2·°C. The solution has a boiling point elevation of 21°C. The heat of dilution is negligible as are radiation losses. Assume that the enthalpy of the solution at all stages equals the enthalpy of pure water at the same temperature.

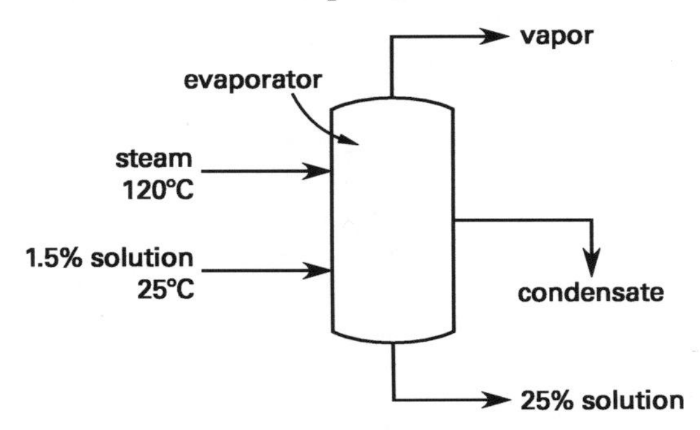

The heat transfer area is most nearly

(A) 0.10 m^2

(B) 64 m^2

(C) 100 m^2

(D) 360 000 m^2

39. Combustion gases leaving a combustion chamber are used in a crossflow heat exchanger as shown. The heat exchanger heats 250 kg/min of process water from a temperature of 28°C to 76°C. During this process, the combustion gases changes from 600°C to 100°C.

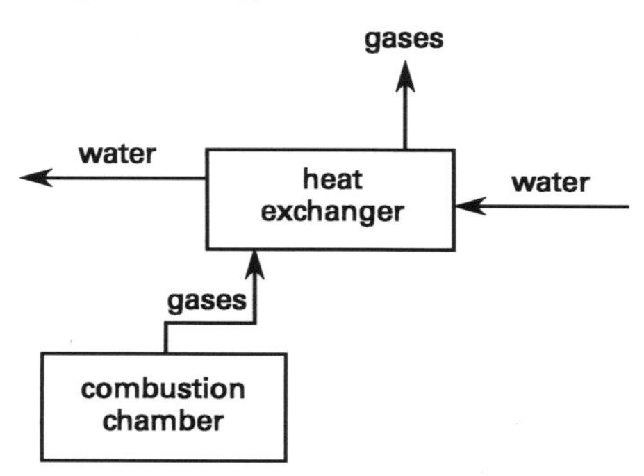

The combustion gases consist of 8.3% (by mole) carbon dioxide (CO_2), 17.18% (by mole) water (H_2O), 69.89% (by mole) nitrogen (N_2) and the balance oxygen (O_2). The molar enthalpies of these component gases at 100°C and at 600°C are

component	enthalpy at 100°C (kJ/mol)	enthalpy at 600°C (kJ/mol)
CO_2	2.90	26.53
H_2O	2.54	20.91
N_2	2.19	17.39
O_2	2.24	18.41

The heat capacity of the process water is 4.184 J/g·°C. The molar flow rate of the combustion gases in the heat exchanger is most nearly

(A) 7.3×10^2 mol/min

(B) 3.0×10^3 mol/min

(C) 3.2×10^3 mol/min

(D) 3.0×10^6 mol/min

40. A vacuum evaporative crystallizer operates at 212°C. The crystallizer receives 4800 lbm of 25% (by weight) aqueous (in H_2O) ammonium sulfate, $(NH_4)_2SO_4$. At this temperature, the stable solid phase is the monohydrate $(NH_4)_2SO_4{\cdot}H_2O$, with a solubility of 46%. The molecular weight of water is 18 lbm/lb mole and that of $(NH_4)_2SO_4$ is 132 lbm/lb mole. During the evaporation, 75% of the water is removed.

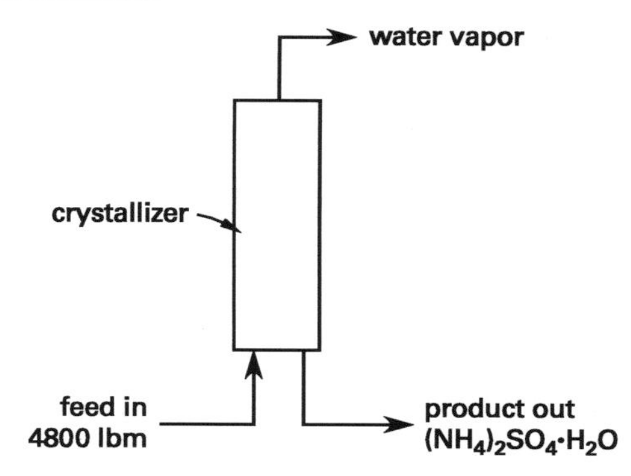

The monohydrate crystals produced have a mass that is most nearly

(A) 420 lbm

(B) 490 lbm

(C) 560 lbm

(D) 2000 lbm

STOP!

DO NOT CONTINUE!

This concludes the Morning Session of the examination. If you finish early, check your work and make sure that you have followed all instructions. After checking your answers, submit your solutions and leave the examination room. Once your answers are submitted you will not be able to access them again.

Afternoon Session Answer Sheet

41. (A) (B) (C) (D)	51. (A) (B) (C) (D)	61. (A) (B) (C) (D)	71. (A) (B) (C) (D)
42. (A)(B)(C)(D)(E)(F)	52. (A) (B) (C) (D)	62. (A) (B) (C) (D)	72. (A) (B) (C) (D)
43. (A) (B) (C) (D)	53. (A) (B) (C) (D)	63. (A) (B) (C) (D)	73. (A) (B) (C) (D)
44. (A) (B) (C) (D)	54. (A) (B) (C) (D)	64. (A) (B) (C) (D)	74. (A) (B) (C) (D)
45. (A) (B) (C) (D)	55. (A) (B) (C) (D)	65. (A) (B) (C) (D)	75. (A) (B) (C) (D)
46. (A) (B) (C) (D)	56. (A) (B) (C) (D)	66. (A) (B) (C) (D)	76. (A) (B) (C) (D)
47. (A) (B) (C) (D)	57. (A) (B) (C) (D)	67. (A) (B) (C) (D)	77. (A) (B) (C) (D)
48. (A) (B) (C) (D)	58. (A) (B) (C) (D)	68. (A) (B) (C) (D)	78. (A) (B) (C) (D)
49. (A) (B) (C) (D)	59. (A) (B) (C) (D)	69. (A) (B) (C) (D)	79. (A) (B) (C) (D)
50. (A) (B) (C) (D)	60. drag and drop	70. (A) (B) (C) (D)	80. (A)(B)(C)(D)(E)(F)

Afternoon Session

41. A tank with a nominal capacity of 2000 gal contains 1000 gal of 12% (by mass) sodium chloride (NaCl) solution in water. An aqueous solution of 21% (by mass) NaCl is fed to the tank at a rate of 42 gal/min starting at time zero. The density of the NaCl solution is constant at 62.4 lbm/ft^3. The contents of the tank are well mixed at all times. After 10 min, the concentration of the NaCl solution in the tank is most nearly

(A) 0.15 lbm/gal

(B) 1.2 lbm/gal

(C) 2.4 lbm/gal

(D) 9.2 lbm/gal

42. The walls of the gas or vapor space and other tank components that are above the maximum liquid level at the top of the tank shall be designed for which of the following?

Select **all** that apply.

(A) a pressure not less than that at which the pressure relief valves are to be set

(B) the maximum positive gauge pressure for which this space is designed shall be understood to be the nominal pressure rating for the tank

(C) the minimum partial vacuum will be greater than that at which the vacuum relief valves are set to open

(D) a maximum partial vacuum that can be developed in the space when the inflow of air through the vacuum relief valves is at its maximum specified rate

(E) pressure increases due to temperature or gravity of the liquid contents of the tank are not to be considered

(F) a pressure not to exceed 15 lbf/in^2 gauge

43. A pipe with an inside diameter of 2.067 in carries carbon dioxide (CO_2) gas. At one point, the CO_2 passes through a square-edged orifice with an inside diameter of 1.00 in. Taps are located 1 diameter upstream and 0.5 diameter downstream from the orifice's inlet face. The differential pressure measured between these two taps is 3 lbf/in^2. The orifice's discharge coefficient is 1.00. The CO_2 has a molecular weight of 44 lbm/lb mole and a ratio of specific heats of 1.29. The gas is under a pressure of 54.7 lbf/in^2 and is 50°F. The flow rate of the CO_2 in the pipe is most nearly

(A) 0.62 lbm/hr

(B) 180 lbm/hr

(C) 390 lbm/hr

(D) 2200 lbm/hr

44. A water solution containing 0.675% (by weight) magnesium sulfate is distilled in a train consisting of two distillation columns as shown.

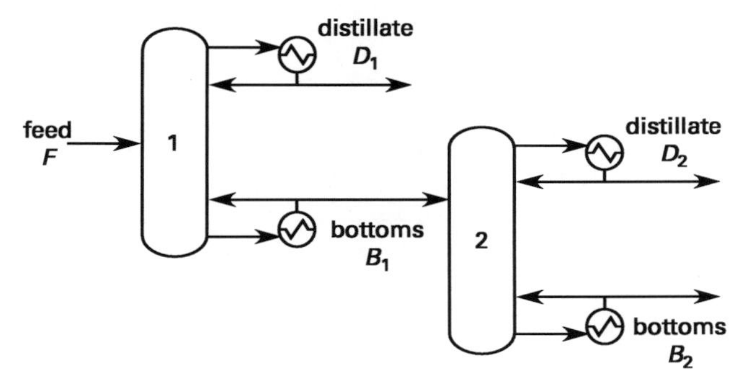

The overhead produced in the first distillation column is pure water at a rate of 1000 lbm/hr. The bottoms stream from the first distillation column is fed into the second distillation column. The overhead produced in the second distillation column is pure water at a rate of 1200 lbm/hr.

To meet specifications, the mass fraction of magnesium sulfate produced in the second distillation column must not be greater than 0.015. To achieve this, the minimum mass flow rate of the feed to the first distillation column is most nearly

(A) 2800 lbm/hr

(B) 3000 lbm/hr

(C) 4000 lbm/hr

(D) 5200 lbm/hr

45. The second-order gas reaction M $\rightarrow$ products is carried out isothermally in a 25 m^3 constant-volume batch reactor. The reactor, which is well mixed, is fed with 32 kmol of M. The reaction rate constant is

0.96 m^3/kmol·h. The conversion of M is 94%. The time necessary to achieve the conversion of M is most nearly

(A) 0.079 h

(B) 0.51 h

(C) 2.9 h

(D) 13 h

46. A solution consists of a solid solute A dissolved in a solvent C. The solution contains 30% A (by weight). 100 lbm of this solution is fed into an extractor, and A is to be extracted one time using a pure solvent S. The extraction is isothermal at 25°C. The raffinate to be extracted is to contain 10% A on a C-free basis. Use the following distribution diagram.

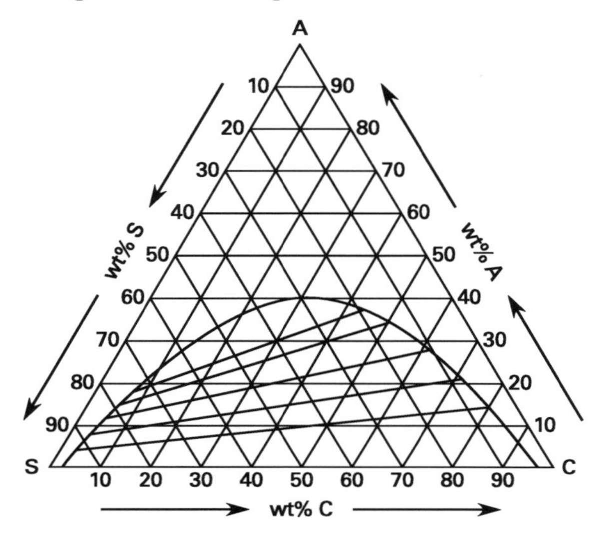

The mass of solvent used in this extraction is most nearly

(A) 6.9 lbm

(B) 150 lbm

(C) 190 lbm

(D) 2200 lbm

47. As ambient air at a pressure of 14.7 lbf/in^2 passes through an adiabatic saturator, its temperature drops from 75°F to 50°F. The average heat capacity of dry air is constant at 0.24 Btu/lbm-°F. The relative humidity of the ambient air is most nearly

(A) 0.11%

(B) 4.0%

(C) 12%

(D) 27%

48. Water flows from one reservoir to another as shown. The water flows at 0.045 m^3/s through a wrought-iron pipe with an inside diameter of 10 cm and a specific roughness of 0.047 mm. The kinematic viscosity of the water is 1.007×10^{-6} m^2/s.

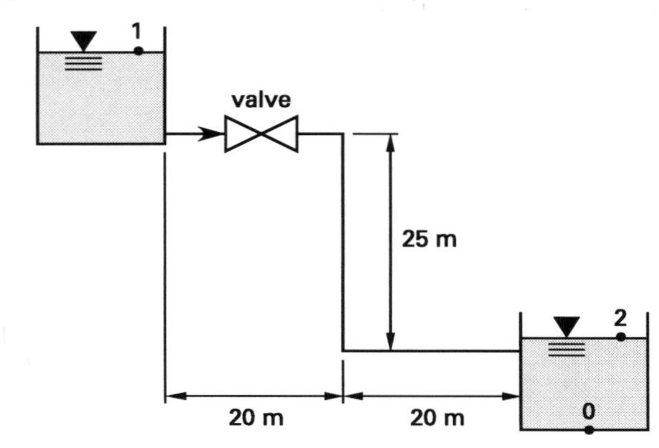

The pipe system includes two elbows, each with a resistance coefficient of 0.64, and one globe valve fully open with a resistance coefficient of 5.7. The resistance coefficient for the entrance is 0.5, and that for the exit is 1.0. The difference in elevation between the two reservoir surfaces is most nearly

(A) 26 m

(B) 33 m

(C) 52 m

(D) 320 m

49. A plug-flow reactor receives a feed consisting of 660,430 gal/day of a wastewater stream containing phenol (C_6H_5OH) with a concentration of 0.00104 lbm/gal. The C_6H_5OH is oxidized in the reactor. The oxidation is a first-order reaction with a reaction constant of 465 day^{-1}. The resulting product concentration is 0.000015 lbm/gal. Assume steady-state reaction. The plug-flow reactor volume is most nearly

(A) 2.0×10^{-4} gal

(B) 3.0×10^{-3} gal

(C) 6.0×10^{3} gal

(D) 1.0×10^{5} gal

50. Three continuous stirred tank reactors (CSTRs), each with a volume of 500 m^3, are arranged in series and initially filled with inert material. The second-order liquid-phase reaction $M + N \rightarrow$ products is carried out using this arrangement. The species M and N are fed in separate lines to the first CSTR, each at a volumetric flow rate of 20 m^3/min and an initial concentration of

3.5 kmol/m^3. The reaction rate constant is 0.078 m^3/kmol·min.

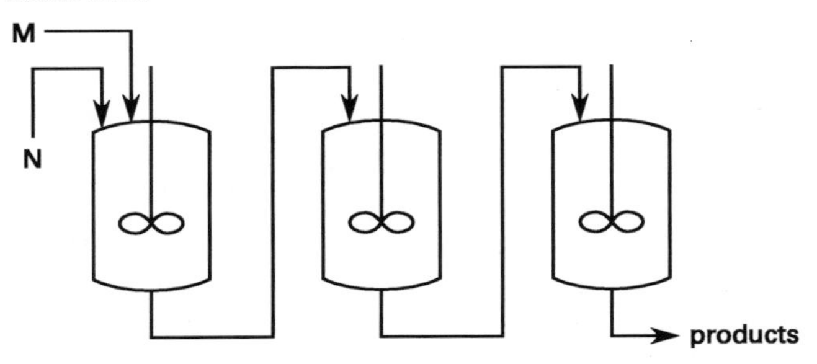

At steady-state conditions, the fractional conversion of M to products is most nearly

(A) 0.30

(B) 0.40

(C) 0.50

(D) 0.80

51. Fuel gas and air are fed into an adiabatic combustion chamber as shown.

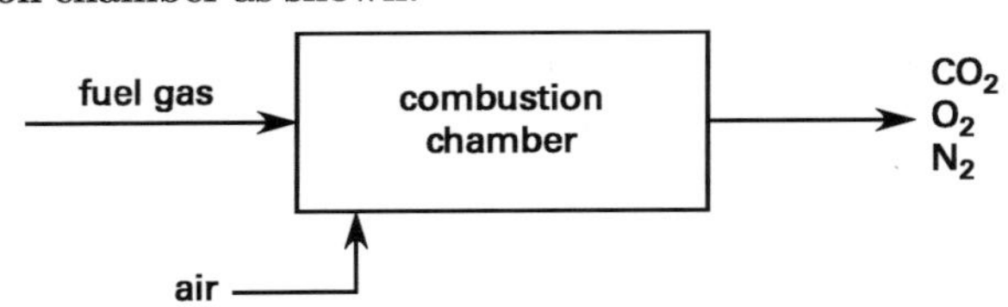

The fuel gas consists of 25% (by mole) carbon monoxide (CO) and 75% (by mole) nitrogen (N_2). 160% excess air is used. Assume that the air is 21% oxygen and 79% nitrogen, with other components negligible. The fuel gas and air both enter the combustion chamber at 25°C. The combustion reaction is complete, and the combustion products consist of carbon dioxide (CO_2), oxygen (O_2), and nitrogen (N_2). The constant average heat capacities for the combustion products are

component	average heat capacity (cal/mol·°C)
CO_2	12.00
O_2	7.90
N_2	7.55

At 25°C, the enthalpy of formation is −94.052 kcal/mol for CO_2 and −26.412 kcal/mol for CO. The flame temperature of the gas is most nearly

(A) 650°C

(B) 850°C

(C) 870°C

(D) 890°C

52. Phosphine, PH_3, can be decomposed into phosphorus, P_4, and hydrogen, H_2. The rate constant, k, for the decomposition has been determined at different temperatures as shown.

k (s^{-1})	T (K)
0.00476	850
0.0620	948
0.249	1000
9.97	1200
299	1500

For the decomposition of PH_3, the activation energy is most nearly

(A) −50 000 cal/mol

(B) 50 000 cal/mol

(C) 143 000 cal/mol

(D) 145 000 cal/mol

53. A mixture of 47% (by mole) benzene (C_6H_6) and 53% (by mole) toluene (C_7H_8) is fed into a flash separator at a rate of 1 mol/h as shown.

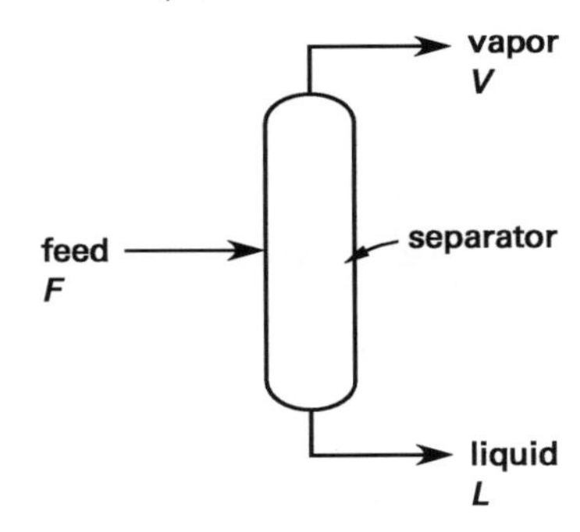

The separator is kept at a constant temperature and pressure. At this temperature and pressure, the ratio of the mole fraction of the benzene in the vapor phase to the mole fraction of the benzene in the liquid phase is 1.748; the ratio of the mole fraction of toluene in the vapor phase to the mole fraction of the toluene in the liquid phase is 0.6223. The ratio of the molar vapor flow rate to the molar liquid flow rate is most nearly

(A) 0.36

(B) 0.89

(C) 1.2

(D) 3.4

54. Over a period of 30 days, the inflow to a lake is 1.51 m^3/s. During this period, precipitation is 7.45 cm, and the lake's volume of stored water increases by an estimated 750 000 m^3. The lake has a surface area of

0.608 km^2. A dam regulates the discharge from the lake to 1.22 m^3/s. The estimated evaporation depth for this period is

(A) 0.077 cm

(B) 7.7 cm

(C) 130 cm

(D) 530 cm

55. A thin film of ethyl acetate at 30°C slides slowly down the inner surface of a wetted-wall tower as shown. Dry air flowing up through the tube is used to remove the film. The dry air flows at a temperature of 30°C and a rate of 0.015 kg/min. The tube has an internal diameter of 10 mm. At 30°C, the diffusivity of ethyl acetate in air is 8.9×10^{-6} m^2/s. The absolute viscosity of air is 1.846×10^{-5} $N \cdot s/m^2$, and the kinematic viscosity is 1.67×10^{-5} m^2/s.

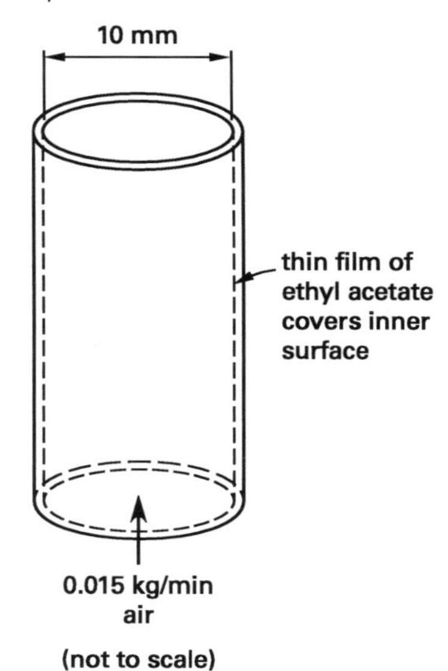

The average mass transfer convection coefficient is most nearly

(A) 5.3×10^{-6} m/s

(B) 5.3×10^{-3} m/s

(C) 2.2×10^{-2} m/s

(D) 5.4×10^{-2} m/s

56. Water flows at 62 gal/min from a supply tank, through a pipe to a pressure vessel operating at 18 lbf/in^2 gauge. The water has a density of 62.2 lbm/ft^3 and a viscosity of 0.86 cP. The horizontal portion of the pipe is at ground level and has an equivalent length of 225 ft. The system includes a surge tank, vented to the atmosphere, as shown. The pipe has a roughness of 0.0002 ft and an inside diameter of 2.469 in. During normal operations, the check valve is fully open.

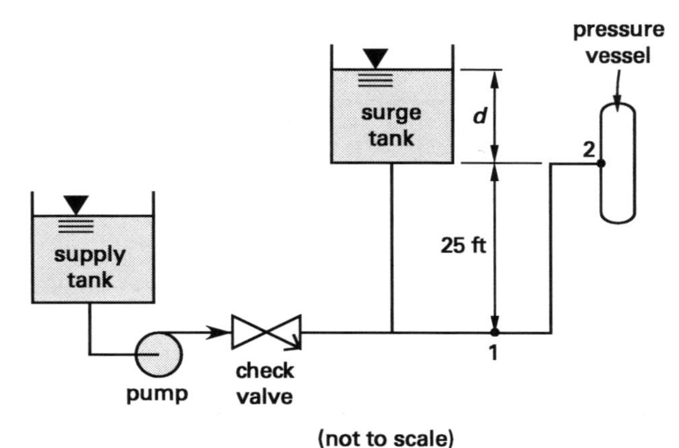

Under normal operating conditions, the depth of the water in the surge tank is most nearly

(A) 6.5 ft

(B) 48 ft

(C) 73 ft

(D) 98 ft

57. Every day, 2300 lbm of waste-activated sludge (dry basis) is treated for stabilization in an aerobic digester with air. The air contains 23.3% oxygen by weight and enters at 72°F and 14.7 lbf/in^2. The solids in the incoming sludge consist of 82% volatiles. These solids consume 2.0 lbm of oxygen per lbm of solid. At least a 63% reduction in solids is needed. Assume the air follows the ideal gas law. The volume of air needed for proper operation of the aerobic digester is most nearly

(A) 10,000 ft^3/day

(B) 32,000 ft^3/day

(C) 140,000 ft^3/day

(D) 350,000 ft^3/day

58. A large open water tank feeds a piping system by gravity. The inside diameter of the pipe is 6.065 in. A branched piping system is installed as shown. Points 1, 2, 3, and 4 lie in the same plane.

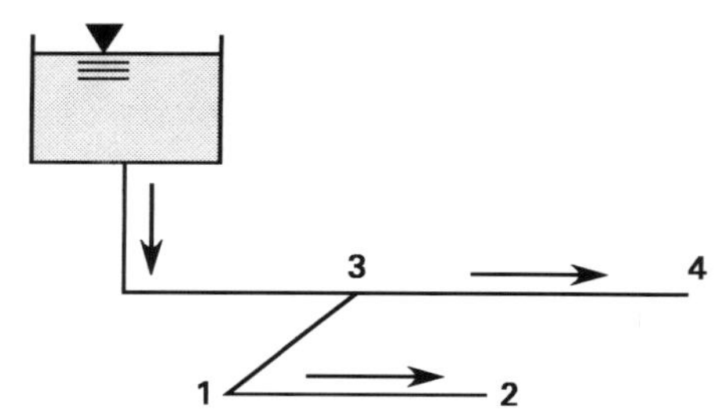

The total equivalent length of pipe from the exit of the tank to point 3 is 2000 ft. The total equivalent lengths of pipes 3-4 and 3-1-2 are 650 ft each. The flow rate at point 2 is 350 gal/min, and the flow rate at point 4 is 1590 gal/min. The density of the water is 62.3 lbm/ft^3, and its viscosity is 1.13 cP. The Darcy friction factor expressed in terms of the Reynolds number is $f = 0.00357 + 0.0218\ \text{Re}^{-0.55}$. The total head loss due to friction is most nearly

(A) 28 ft

(B) 76 ft

(C) 100 ft

(D) 130 ft

59. A mixer receives two streams as shown, one consisting of pure ethylene oxide (C_2H_4O) and flowing at a rate of 1 mol/h, and the other consisting of pure water (H_2O) and flowing at a rate of 4.8 mol/h.

The outlet from the mixer is fed to a heater. The feed enters the heater at 30°C and is heated to 86°C. In the heating process, the pressure is kept constant at 1 bar.

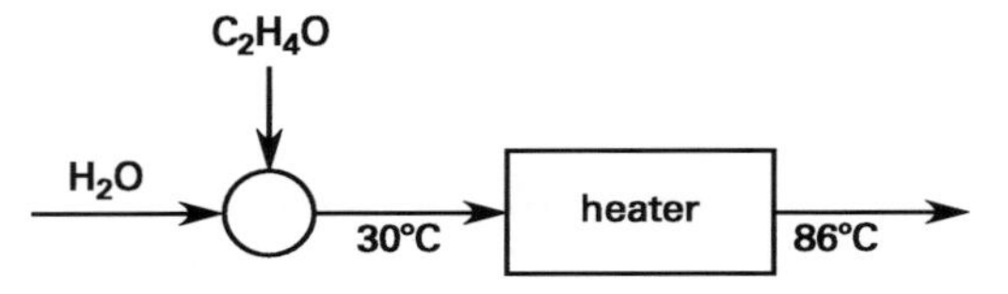

The heat capacity of the C_2H_4O as a function of absolute temperature is

$$\begin{aligned} c_{p,C_2H_4O} &= -3.2032 \times 10^{-3}\ \frac{\text{kJ}}{\text{mol·K}} \\ &\quad + \left(1.9521 \times 10^{-4}\ \frac{\text{kJ}}{\text{mol·K}^2}\right)T \\ &\quad - \left(7.7343 \times 10^{-8}\ \frac{\text{kJ}}{\text{mol·K}^3}\right)T^2 \end{aligned}$$

Similarly, the heat capacity of the water is

$$\begin{aligned} c_{p,H_2O} &= 72.84 \times 10^{-3}\ \frac{\text{kJ}}{\text{mol·K}} \\ &\quad + \left(1.0400 \times 10^{-5}\ \frac{\text{kJ}}{\text{mol·K}^2}\right)T \\ &\quad - \left(1.4976 \times 10^{-9}\ \frac{\text{kJ}}{\text{mol·K}^3}\right)T^2 \end{aligned}$$

The rate of heat added to the mixture in the heater is most nearly

(A) 3.0 kJ/h

(B) 18 kJ/h

(C) 20 kJ/h

(D) 23 kJ/h

60. A list of symbols denoting different containers and vessels, and a list of corresponding labels, is shown.

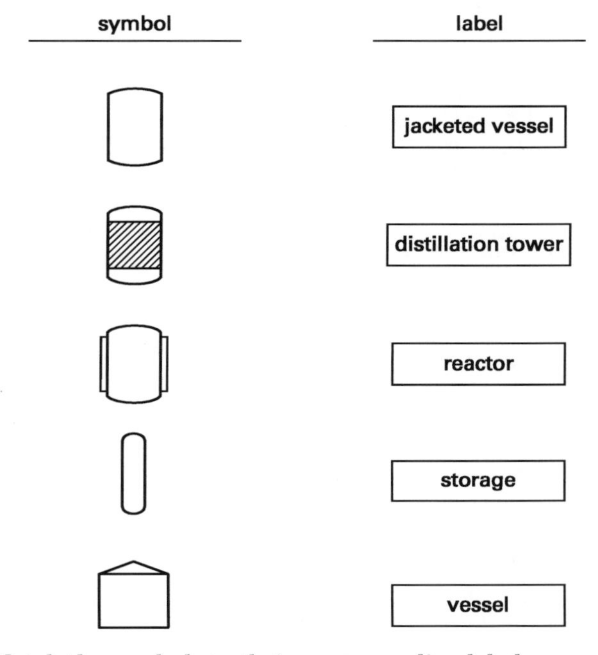

Match the symbols to their corresponding labels.

61. The gaseous phase elementary reaction $M \rightarrow N + 2P$ is carried out in a plug-flow reactor (PFR) at 260°F and 118 lbf/in^2. The PFR is fed with pure M at 302°F at a rate of 25 lb mole/min. The reaction is irreversible, and a conversion of 85% is achieved. The reaction rate constant at 140°F is 0.002 min^{-1}. The activation energy for the reaction is 37,188.31 Btu/lb mole.

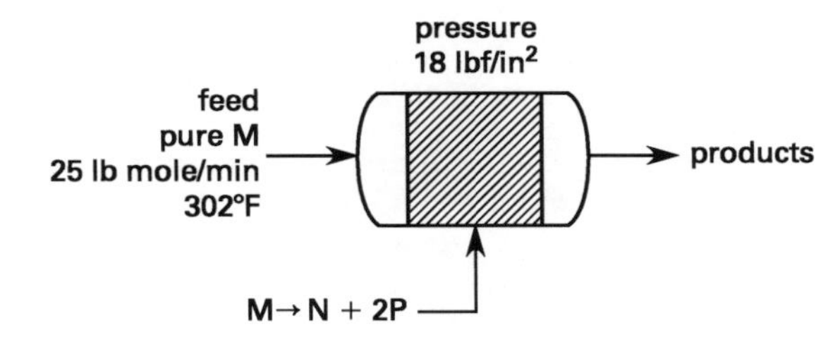

The PFR volume needed to achieve the desired conversion is most nearly

(A) 2.7×10^3 ft^3

(B) 1.8×10^4 ft^3

(C) 2.9×10^6 ft^3

(D) 3.3×10^6 ft^3

62. Saturated water at a pressure of 2200 lbf/in^2 enters a single seat control valve. The valve operates at a back pressure of 1200 lbf/in^2. It has been determined that when pure water at 60°F flows through the valve at a flow rate of 425 gal/min, the pressure drop through the valve is 1 lbf/in^2. The saturated water has a specific volume of 0.0267 ft^3/lbm. At the entrance of the valve, the enthalpy of the saturated water is 695 Btu/lbm. At the exit of the valve, the specific enthalpy of the liquid is 572 Btu/lbm and the specific enthalpy of the vapor is 1180 Btu/lbm. The density of the water is 62.4 lbm/ft^3. The mass flow rate through the valve, corrected for flashing, is most nearly

(A) 18,000 lbm/min

(B) 29,000 lbm/min

(C) 69,000 lbm/min

(D) 87,000 lbm/min

63. A vent stream enters a condenser as shown.

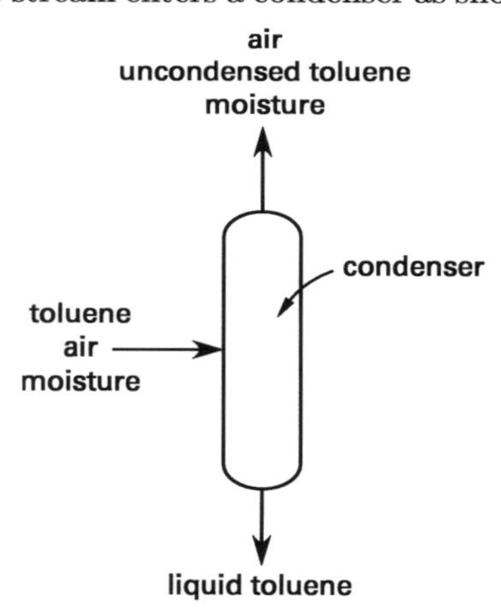

The vent stream is at 150°F and consists of air, 37.5% (by mole) toluene, and a negligible amount of moisture. The stream is fed at a rate of 100 ft^3/min. The condenser operates at a pressure of 760 mm Hg. The average heat capacity of toluene is 24.77 Btu/lb mole-°F and that of air is 6.95 Btu/lb mole-°F. The heat of condensation of toluene is 14,395 Btu/lb mole. The relationship between the condensation temperature of toluene in degrees Fahrenheit and the partial pressure of toluene in millimeters of mercury is

$$T_{\text{cond},°\text{F}} = \left(\frac{1344.8}{6.955 - \log P_{\text{toluene}}} - 219.48 \right)(1.8°\text{F}) + 32°\text{F}$$

The condenser-required removal efficiency of toluene is 92%. The condenser heat load needed to meet the required removal efficiency of toluene is most nearly

(A) 4.5×10^3 Btu/hr

(B) 7.5×10^4 Btu/hr

(C) 7.8×10^4 Btu/hr

(D) 7.9×10^4 Btu/hr

64. Air conditioning equipment is being designed to serve two processing rooms with different needs. The first room requires 134 m^3/min of air at 24°C and 28% relative humidity. The second room requires 65 m^3/min of air at 32°C and 53% relative humidity.

These two airstreams will be produced by feeding fresh air into a cooler/condenser system. Part of the cool stream from the cooler/condenser system will then be passed through a heater to provide the supply air to the first room. The balance of the cool air will be mixed with fresh air and heated in a second heater to provide the air needed in the second room.

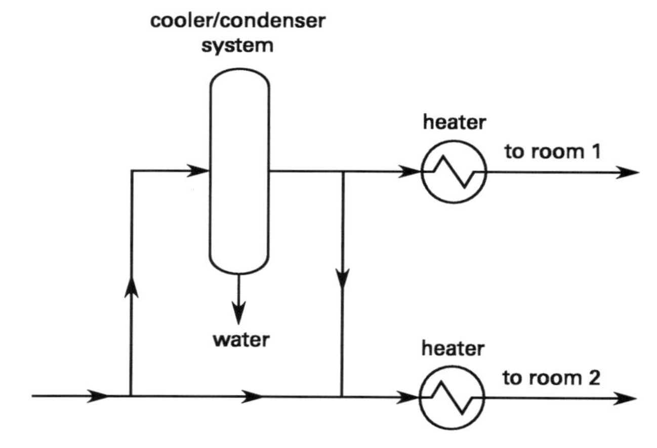

The available fresh air supply has a dry-bulb temperature of 35°C and a wet-bulb temperature of 25°C. The mass flow rate at which the water in the cooler/condenser must be removed is most nearly

(A) 0.78 kg/min

(B) 1.4 kg/min

(C) 1.7 kg/min

(D) 5.7 kg/min

65. Water flows through a thin-walled pipe at 27 kg/min. The pipe is 50 mm in diameter and 10 m long.

Steam condenses on the outer surface of the pipe, which maintains a uniform surface temperature of 90°C. The water is 20°C at the inlet and 68°C at the outlet. The average heat capacity of the water is constant at 4280 J/kg·K. The average convection coefficient associated with the water flow is most nearly

(A) 1.4 W/m²·K

(B) 1400 W/m²·K

(C) 59 000 W/m²·K

(D) 85 000 W/m²·K

66. Liquid *n*-butane (n-C_4H_{10}) is isomerized to produce liquid isobutane (i-C_4H_{10}) in an adiabatic plug-flow reactor (PFR) as shown.

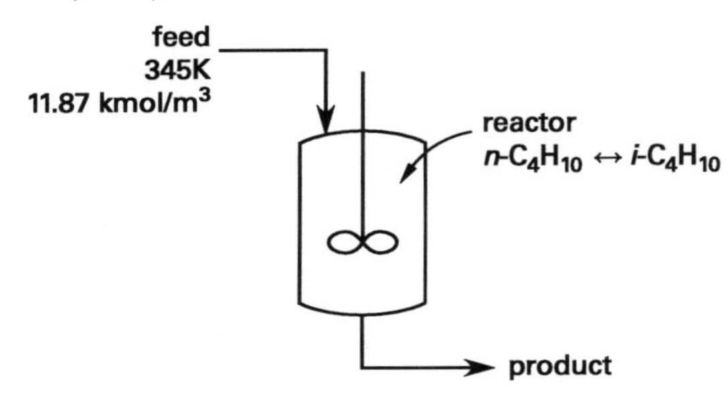

The reaction is first order and reversible. The specific reaction rate constant is 62.2 h^{-1} at a temperature of 380K. The reaction is exothermic with a heat of reaction of 6900 J/mol of *n*-butane. The PFR is fed with 400 kmol/h of a mixture containing 86% *n*-butane and 14% inert ingredients (by mole). The heat capacity of *n*-butane is 141 J/mol·K, and that of the inert portion is 161 J/mol·K. The feed is at a temperature of 345K and an initial concentration of 11.87 kmol/m³. The isomerization reaction has an equilibrium constant of 3.09 at a temperature of 335K and activation energy of 65.7 kJ/mol. To achieve a conversion of 55% of *n*-butane, the required volume of the PFR is most nearly

(A) 0.29 m³

(B) 1.8 m³

(C) 2.2 m³

(D) 4.0 m³

67. The insulation system of a furnace wall is constructed using refractory brick in the inner wall, firebrick in the middle, and insulating firebrick in the outer wall. Under steady-state conditions, the inner and outer surface temperatures are 860°C and 20°C, respectively. The oven air temperature is 1500°C, and the inside convection coefficient is 25 W/m²·K. The thickness and known thermal conductivities of the wall materials are as shown.

material	thickness, δ	thermal conductivity, k
refractory brick	0.30 m	45 W/m·K
firebrick	0.15 m	?
insulating brick	0.20 m	20 W/m·K

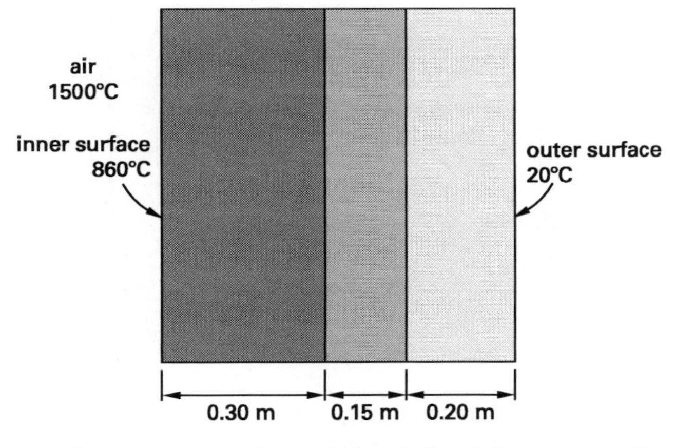

The thermal conductivity of the firebrick is most nearly

(A) 2.0 W/m·K

(B) 2.7 W/m·K

(C) 4.2 W/m·K

(D) 25 W/m·K

68. A feed stream of combustion gas and a stream of air enter a combustion chamber as shown. The combustion products are water and a flue gas.

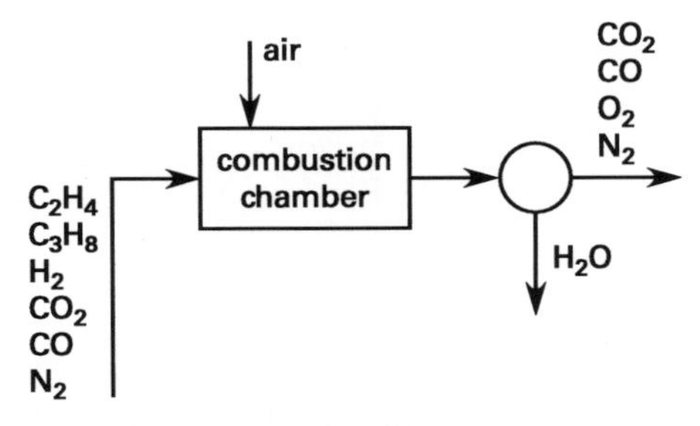

The combustion gas has the following composition (in percentages by mole).

component	mole %
C_2H_4	15.10
C_3H_8	14.20
H_2	35.40
CO_2	28.10
CO	2.90
N_2	4.30

After combustion, the composition of the flue gas (in percentages by mole) is

component	mole %
CO_2	11.50
CO	0.35
O_2	5.20
N_2	82.90
other gases (negligible)	0.05

Assume the air is 79% (by mole) nitrogen and 21% (by mole) oxygen and that other components are negligible. The number of moles of air used in the combustion is most nearly

(A) 5 lb mole

(B) 6 lb mole

(C) 7 lb mole

(D) 9 lb mole

69. The wall of a furnace is constructed from a material with a thermal conductivity of 1.6 W/m·K. The wall is 0.2 m thick and has an emissivity of 0.85. The wall's outer surface is at 126°C, and it touches still air at 23°C. The free convection heat transfer coefficient is 25 W/m^2·K.

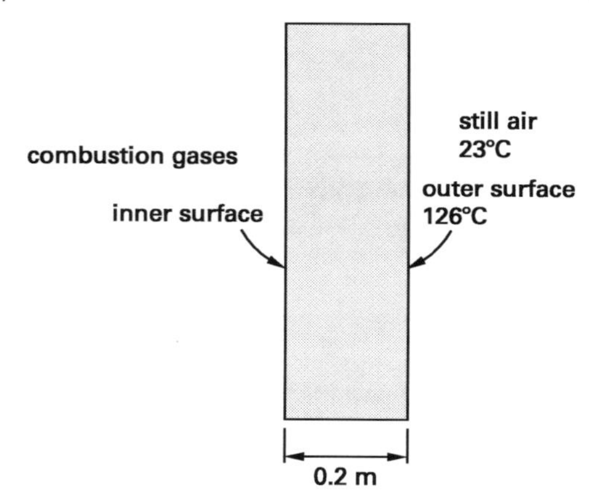

When the wall has reached steady-state conditions, the temperature of the wall surface that is in contact with the combustion gases is most nearly

(A) 340°C

(B) 450°C

(C) 550°C

(D) 830°C

70. A plant is being designed to operate 300 day/yr processing 900,000 lbm/day of a liquor of water and caustic soda. The liquor in the feed will contain 6% (by weight) caustic soda, and this must be concentrated by evaporation to 38% in the product. Either a single-effect or a multiple-effect evaporator will be used. A single-effect evaporator with the needed capacity requires an initial investment of $36,000. This same investment is required for each additional effect. The service life is estimated to be 10 years, and the salvage value of each effect at the end of the service life is estimated to be $10,000. Annual fixed charges (other than depreciation) will equal 20% of the initial investment. Steam costs $0.72 per 1000 lbm. The remaining administration, labor, and miscellaneous costs of operating the plant will be $80/day, regardless of how many evaporator effects are used. The number of pounds of steam needed is 1.11 times of the number of pounds of water evaporated in each effect. Use straight-line depreciation to evaluate the alternatives. The number of effects that will give the minimum total cost per day is

(A) 2

(B) 3

(C) 4

(D) 5

71. In the simplified process flow sheet shown, a mixture consisting of (by mole) 16% xylene, 35% toluene, and 49% benzene is separated. The feed stream is introduced into distillation column 1 at a rate of 82 mol/min.

The distillate from column 1 is fed into distillation column 2. The distillate from column 2 consists of 26% xylene, 53% toluene, and 21% benzene. The bottoms from column 2 consist of 8% xylene, 40% toluene, and 52% benzene.

The bottoms from column 1 are fed into distillation column 3. The distillate from column 3 consists of 7% xylene, 42% toluene, and 51% benzene. The bottoms from column 3 consist of 16% xylene, 15% toluene, and 69% benzene. The bottoms stream from column 3 has a molar flow rate of 30 mol/min.

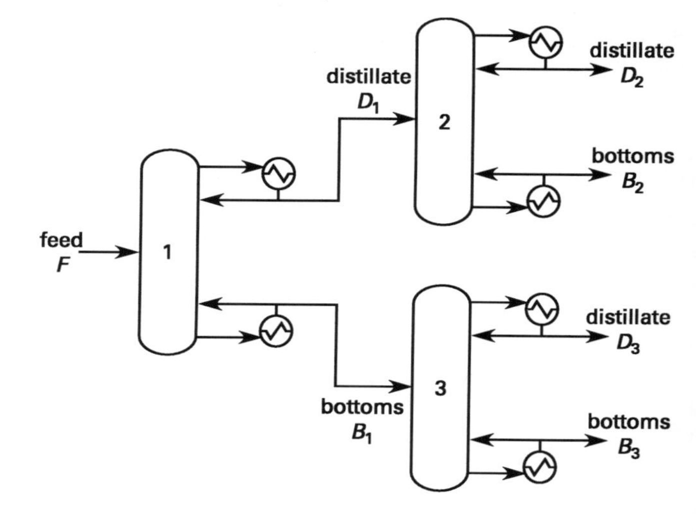

The mole fraction of benzene in the bottoms stream produced in the distillation column 1 is most nearly

(A) 0.34

(B) 0.53

(C) 0.63

(D) 0.75

72. A solution of 48% (by weight) methanol and 52% water at 80°F is to be continuously distilled at a rate of 6000 lbm/hr and a pressure of 1 atm to provide a distillate containing 94% methanol and a residue containing 1.0% methanol.

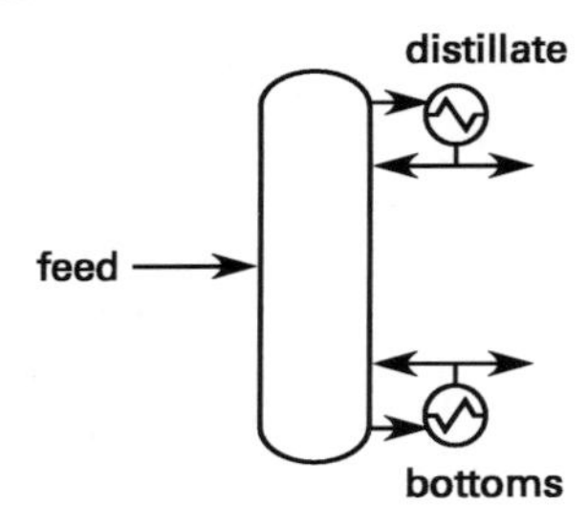

The mass flow rate of the distillate produced is most nearly

(A) 2970 lbm/hr

(B) 3000 lbm/hr

(C) 3200 lbm/hr

(D) 3300 lbm/hr

73. A shell-and-tube heat exchanger used in a steam power plant consists of a single shell with 10 000 U-tubes, each performing two passes. The tubes are of thin-wall construction with a diameter of 50 mm. On the outer surface of the tubes, the steam condenses to a saturated liquid at a temperature of 60°C and with a convection coefficient of 1223 $W/m^2{\cdot}K$. The cooling fluid used is water flowing in the tubes. The cooling water enters the heat exchanger at a rate of 36 000 kg/s and a temperature of 25°C. The viscosity of the cooling water is 0.0009 $N{\cdot}s/m^2$. The heat transfer rate in the exchanger is 8.53×10^8 W. The average specific heat capacity of the cooling water is 4183 J/kg·K. The thermal conductivity of the cooling water is 0.61 W/m·K, and the Prandtl number is 5.90.

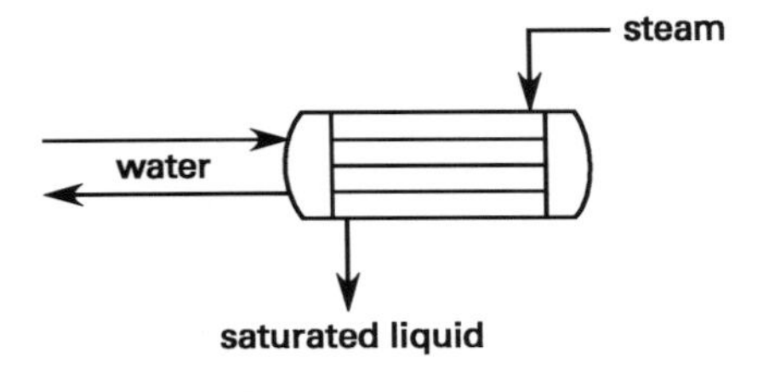

The tube length needed per pass to achieve the required heat transfer is most nearly

(A) 1.5 m

(B) 6.9 m

(C) 8.4 m

(D) 17 m

74. A solid feed consisting of 1675 kg/h of sodium sulfate (Na_2SO_4) and 891 kg/h of an inert, insoluble material is washed in three countercurrent stages to recover Na_2SO_4. Pure water, at a rate of 3880 kg/h, is used as the solvent. The liquid overflow from each stage consists of a solution of Na_2SO_4 dissolved in water and contains none of the inert material. The underflow from each stage is a slurry consisting of the inert material wetted by the Na_2SO_4 solution. The mass of the solution is a constant 52% of the total mass of the slurry. Assume that all the Na_2SO_4 dissolves in the water and that each stage reaches equilibrium. At the end of the third stage, the fraction recovery of Na_2SO_4 is most nearly

(A) 0.027

(B) 0.046

(C) 0.57

(D) 0.99

75. A fuel gas consists of 61% (by mole) methane (CH_4), 36% (by mole) propane (C_3H_8), and the rest nitrogen (N_2). This fuel gas is burned with 25% (by mole) excess air in a combustion chamber as shown. All the fuel is consumed and the combustion is complete. The combustion products comprise carbon dioxide (CO_2), oxygen (O_2), nitrogen (N_2), and water (H_2O).

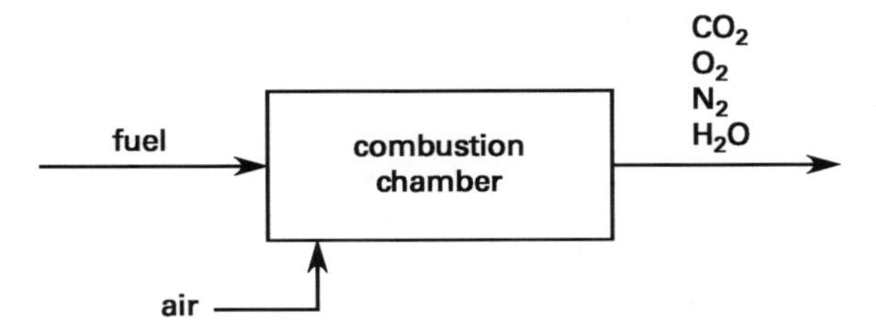

The air fed to the combustion chamber is at 36°C dry-bulb temperature and 32°C wet-bulb temperature. Assume that air is (by mole) 79% nitrogen and 21% oxygen and that other components are negligible. The molecular weight of water is 18 g/mol.

For each mole of fuel gas fed into the combustion chamber, the total number of moles in the outlet stream is most nearly

(A) 5.9 mol

(B) 17 mol

(C) 19 mol

(D) 20 mol

76. A centrifugal pump draws an aqueous solution at 180 gal/min from an open tank vented to the atmosphere as shown.

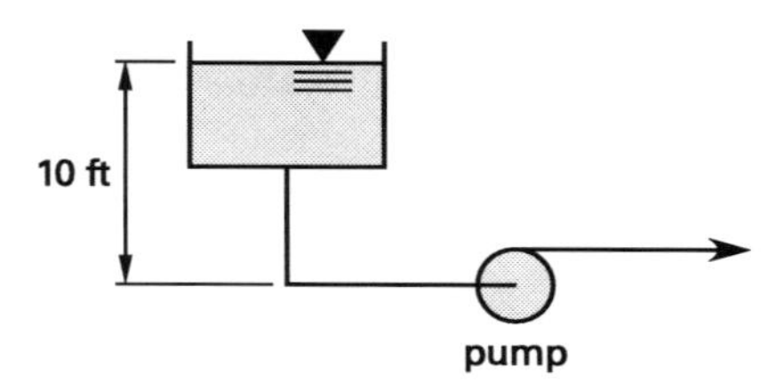

The liquid level in the tank is 10 ft above the entrance to the pump. The equivalent length of the pump suction line is 52.26 ft of 4 in schedule-40 pipe. The roughness of the pipe is 0.00015 ft. The fluid has a density of 60.58 lbm/ft^3, a viscosity of 0.35 cP, and a vapor pressure of 7.51 lbf/in^2. The resistance coefficient for the sudden contraction is 0.4. The net positive suction head available for the specified flow is most nearly

(A) 18 ft

(B) 26 ft

(C) 27 ft

(D) 44 ft

77. A horizontal pipe carrying high-pressure steam passes through a large room. The pipe has an outside diameter of 0.1 m and an outside surface temperature of 234°C; the pipe's emissivity is 0.80. The walls of the room are at a constant 20°C. At the conditions of this flow, the average Nusselt number depends on the Rayleigh and Prandtl numbers and can be obtained from the formula

$$\overline{\text{Nu}} = \left(0.60 + \frac{0.387(\text{Ra})^{1/6}}{\left(1 + \left(\frac{0.559}{Pr}\right)^{9/16}\right)^{8/27}} \right)^2$$

The physical properties of the air at 20°C are as follows.

conductivity, k	0.0338 W/m·K
kinematic viscosity, ν	2.64×10^{-5} m^2/s
thermal diffusivity, α	3.83×10^{-5} m^2/s
Prandtl number, Pr	0.697
thermal expansion coefficient, β	2.73×10^{-3} K^{-1}

Assume the pipe's surface area is small compared to its surroundings, and assume that the room air is still. The heat loss from the pipe per unit length is most nearly

(A) 540 W/m

(B) 840 W/m

(C) 950 W/m

(D) 1400 W/m

78. For the distillation diagram shown, the molar flow rates of the following streams are known.

feed	282 kmol/h
distillate	162 kmol/h
reflux	2300 kmol/h

The molar specific enthalpies of the streams are also known.

feed	11 774 kJ/kmol
vapor	23 912 kJ/kmol
distillate	11 753 kJ/kmol
reflux	11 753 kJ/kmol
bottoms	16 156 kJ/kmol

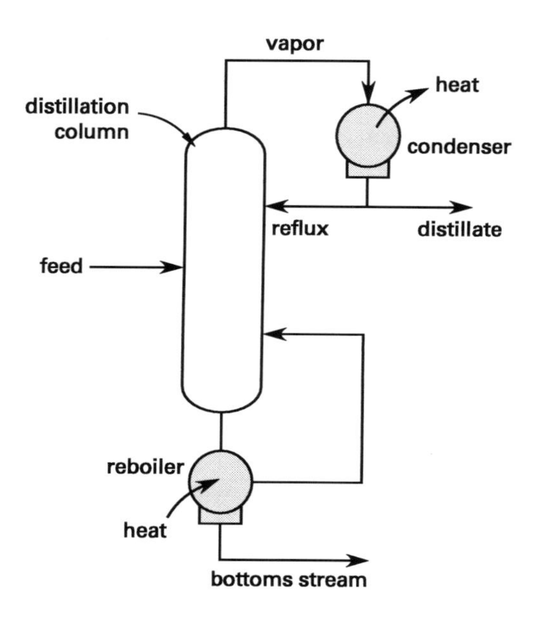

The reboiler duty is most nearly

(A) -3.0×10^7 kJ/h

(B) 5.0×10^5 kJ/h

(C) 3.0×10^7 kJ/h

(D) 4.0×10^7 kJ/h

79. A batch still is loaded with 100 kmol of a liquid containing a binary mixture of 62% (by mole) styrene and 38% toluene. The distillation is carried out at a pressure of 1.2 atm, and the vapor is in equilibrium with the perfectly mixed liquid in the still. An average product composition of 15% toluene is required. The vapor pressure of styrene in millimeters of mercury is given by

$$P_s = e^{14.3284-(3516.43\text{K}/(T-56.1529\text{K}))}$$

T is the bubble point in the absolute scale. Similarly, the vapor pressure of toluene in millimeters of mercury is

$$P_t = e^{14.2515-(3242.38\text{K}/(T-47.1806\text{K}))}$$

The number of moles of charge remaining in the still when the final composition is achieved is most nearly

(A) 0.50 kmol

(B) 4.9 kmol

(C) 16 kmol

(D) 31 kmol

80. Why are rupture disks installed in series with relief valves?

Select **all** that apply.

(A) This configuration gives absolute isolation when handling flammable gases.

(B) This configuration gives absolute isolation when handling extremely toxic chemicals.

(C) The rupture disk burst temperature is lower than the vessel's normal operating temperature.

(D) This configuration protects the relatively complex parts of a spring-loaded device from reactive monomers that could cause plugging.

(E) This configuration protects the vessel against a serious over-pressure incident where water hammer can be experienced.

(F) This configuration protects an expensive spring-loaded device from a corrosive environment.

STOP!

DO NOT CONTINUE!

This concludes the Afternoon Session of the examination. If you finish early, check your work and make sure that you have followed all instructions. After checking your answers, submit your solutions and leave the examination room. Once your answers are submitted you will not be able to access them again.

Answer Key

Morning Session

No.	Answer	No.	Answer	No.	Answer	No.	Answer
1.	D	11.	C	21.	D	31.	C
2.	B	12.	sensor	22.	C	32.	C
3.	A	13.	A	23.	B	33.	see solution
4.	B	14.	loop	24.	C	34.	C
5.	B	15.	B	25.	D	35.	A
6.	B	16.	C	26.	D	36.	D
7.	C	17.	B	27.	see solution	37.	A
8.	B	18.	see solution	28.	D	38.	C
9.	C	19.	C	29.	D	39.	B
10.	B	20.	B	30.	B	40.	C

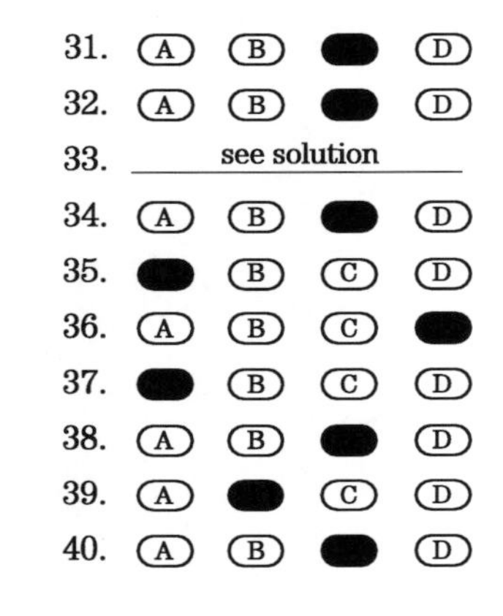

Afternoon Session

No.	Answer	No.	Answer	No.	Answer	No.	Answer
41.	B	51.	D	61.	B	71.	C
42.	A, B, D, F	52.	B	62.	A	72.	B
43.	D	53.	C	63.	D	73.	C
44.	C	54.	B	64.	C	74.	D
45.	D	55.	B	65.	B	75.	D
46.	C	56.	B	66.	C	76.	B
47.	C	57.	C	67.	C	77.	D
48.	B	58.	D	68.	D	78.	C
49.	C	59.	D	69.	C	79.	D
50.	D	60.	see solution	70.	C	80.	A, B, D, F

Solutions

Morning Session

1. The concentration of species M into the plug-flow reactor (PFR) is given as

$$C_{M,0} = 3.25\ \text{kmol/m}^3$$

The space-time is given as

$$\tau = 1.3\ \text{min}$$

From *NCEES Handbook:* Second-Order Reactions, the space-time is

$$k\tau = \frac{C_{Ao} - C_A}{C_A^2}$$

Applying the preceding equation to the problem statement gives

$$k\tau = \frac{C_{M,0} - C_M}{C_M^2} = \frac{\dfrac{C_{M,0}}{C_{M,1}} - 1}{C_{M,0}} = \frac{C_{M,0} - C_{M,1}}{C_{M,0}C_{M,1}}$$

Solving the preceding equation for the concentration of species M in the outlet of the PFR gives

$$\begin{aligned} C_{M,1} &= \frac{1}{k\tau + \dfrac{1}{C_{M,0}}} \\ &= \frac{1}{\left(0.987\ \dfrac{\text{m}^3}{\text{kmol}\cdot\text{min}}\right)(1.3\ \text{min}) + \dfrac{1}{3.25\ \dfrac{\text{kmol}}{\text{m}^3}}} \\ &= 0.629\ \text{kmol/m}^3 \end{aligned}$$

The mass balance of species M around the CSTR gives

$$C_{M,1} = k\tau C_{M,2}^2 + C_{M,2}$$

$$k\tau C_{M,2}^2 + C_{M,2} - C_{M,1} = 0$$

From *NCEES Handbook:* Polynomials, the quadratic formula gives

$$\begin{aligned} C_{M,2} &= \frac{-b \pm \sqrt{b^2 - 4ac}}{2a} \\ &= \frac{-1 \pm \sqrt{1^2 - 4k\tau(-C_{M,1})}}{2k\tau} \\ &= \frac{-1 \pm \sqrt{\begin{array}{l} 1^2 - (4)\left(0.987\ \dfrac{\text{m}^3}{\text{kmol}\cdot\text{min}}\right)(1.3\ \text{min}) \\ \times\left(-0.629\ \dfrac{\text{kmol}}{\text{m}^3}\right) \end{array}}}{(2)\left(0.987\ \dfrac{\text{m}^3}{\text{kmol}\cdot\text{min}}\right)(1.3\ \text{min})} \\ &= 0.4116\ \text{kmol/m}^3 \end{aligned}$$

From *NCEES Handbook:* Reaction Parameters, the fractional conversion is

$$X_A = \frac{C_{Ao} - C_A}{C_{Ao}}$$

Applying the preceding equation to the conditions of the problem statement and solving for the corresponding conversion gives

$$\begin{aligned} X &= \frac{C_{M,0} - C_{M,2}}{C_{M,0}} \\ &= \frac{3.25\ \dfrac{\text{kmol}}{\text{m}^3} - 0.4116\ \dfrac{\text{kmol}}{\text{m}^3}}{3.25\ \dfrac{\text{kmol}}{\text{m}^3}} \\ &= 0.873 \quad (0.90) \end{aligned}$$

The answer is (D).

2. The absolute temperature of the oxygen stream at the inlet is

$$T_{in} = 30°\text{C} + 273° = 303\text{K}$$

The latent heat of the saturated steam, λ, is given as 2230.95 kJ/kg. As the saturated steam condenses, it gives off heat; because the heat exchanger is insulated, the oxygen stream increases in temperature as it receives this heat. The energy given off by the saturated steam during the heat exchange is

$$\dot{Q} = \dot{m}_{\text{steam}}\lambda = \left(225\ \frac{\text{kg}}{\text{h}}\right)\left(2230.95\ \frac{\text{kJ}}{\text{kg}}\right) = 501\,963.75\ \text{kJ/h}$$

From *NCEES Handbook:* Overall Heat-Transfer Coefficient, the energy balance around a heat exchanger is

$$\dot{Q} = \dot{m}c_p(T - T_1)$$

Accordingly, the energy received by the oxygen stream is

$$\dot{Q} = \dot{m}_{\text{oxygen}}\int_{303\text{K}}^{T} c_p dT = \left(125\ \frac{\text{kmol}}{\text{h}}\right) \times \int_{303\text{K}}^{T}\left(29.88\ \frac{\text{kJ}}{\text{kmol}\cdot\text{K}} - \left(0.011\,38\ \frac{\text{kJ}}{\text{kmol}\cdot\text{K}^2}\right)T\right)dT$$

Replacing and integrating,

$$501\,963.75\ \frac{\text{kJ}}{\text{h}} = \left(125\ \frac{\text{kmol}}{\text{h}}\right) \times \left(\begin{array}{l}\left(29.88\ \frac{\text{kJ}}{\text{kmol}\cdot\text{K}}\right)(T - 303\text{K}) \\ -(0.5)\left(0.011\,38\ \frac{\text{kJ}}{\text{kmol}\cdot\text{K}^2}\right) \\ \times\left(T^2 - (303\text{K})^2\right)\end{array}\right)$$

Collecting like terms gives

$$\left(-0.711\,25\ \frac{\text{kJ}}{\text{h}\cdot\text{K}^2}\right)T^2 + \left(3735\ \frac{\text{kJ}}{\text{h}\cdot\text{K}}\right)T - 1\,568\,370\ \frac{\text{kJ}}{\text{h}} = 0\ \text{kJ/h}$$

From *NCEES Handbook:* Polynomials, the roots of a quadratic equation are

$$x_{1,2} = \frac{-b \pm \sqrt{b^2 - 4ac}}{2a}$$

Using the quadratic formula to solve for the outlet temperature of the oxygen stream,

$$T = \frac{-b \pm \sqrt{b^2 - 4ac}}{2a} = \frac{-3735\ \frac{\text{kJ}}{\text{h}\cdot\text{K}} \pm \sqrt{\left(3735\ \frac{\text{kJ}}{\text{h}\cdot\text{K}}\right)^2 - (4)\left(-0.711\,25\ \frac{\text{kJ}}{\text{h}\cdot\text{K}^2}\right) \times\left(-1\,568\,370\ \frac{\text{kJ}}{\text{h}}\right)}}{(2)\left(-0.711\,25\ \frac{\text{kJ}}{\text{h}\cdot\text{K}^2}\right)}$$

There are two solutions to the equation, $T = 460.25\text{K}$ and $T = 4791.1\text{K}$. The latter is unrealistic and can be discarded. Accepting the remaining solution and converting to degrees Celsius,

$$T = 460\text{K} - 273° = 187°\text{C} \quad (190°\text{C})$$

The answer is (B).

3. The relevant data are shown in the illustration.

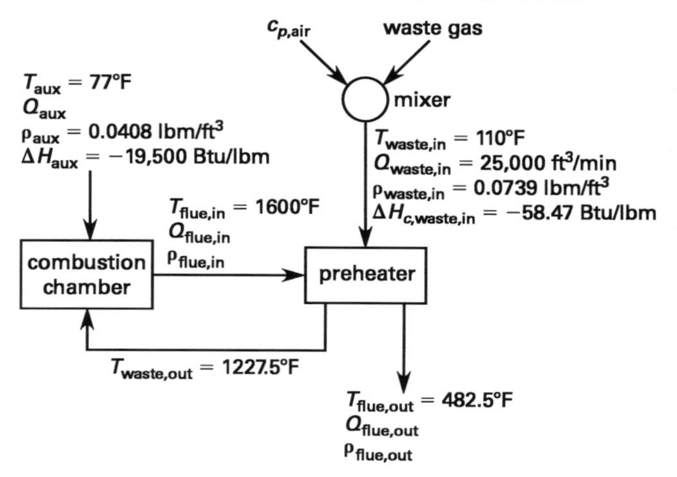

The flow rate of auxiliary fuel is to be measured at a reference temperature, T_{ref}, of 77°F. The density of the waste feed equals the density of the flue gas.

$$\rho_{\text{waste,in}} = \rho_{\text{flue,out}} = 0.0739\ \text{lbm/ft}^3$$

The flow rates of the waste feed entering and leaving the preheater are the same.

$$\dot{V}_{\text{waste,in}} = \dot{V}_{\text{waste,out}} = 25{,}000\ \text{ft}^3/\text{min}$$

The mean heat capacities of the waste and flue gas streams on both sides of the preheater are approximately the same as that of the air.

$$\begin{aligned} c_{p,\text{mean}} &\approx c_{p,\text{waste,in}} \\ &\approx c_{p,\text{flue,in}} \\ &\approx c_{p,\text{flue,out}} \\ &\approx c_{p,\text{air}} \\ &= 0.255 \text{ Btu/lbm-°F} \end{aligned}$$

As in *NCEES Handbook:* Conduction, the rate of heat transferred in or out of a flowing material is

$$Q_{\text{total}} = \rho V c_p (T_i - T)$$

The heat content of the inlet waste feed is

$$\dot{Q}_{\text{waste,in}} = \rho_{\text{waste,in}} \dot{V}_{\text{waste,in}} c_{p,\text{mean}} (T_{\text{waste,in}} - T_{\text{ref}})$$

The heat content of the outlet flue gas is

$$\dot{Q}_{\text{flue,out}} = \rho_{\text{flue,out}} \dot{V}_{\text{flue,out}} c_{p,\text{mean}} (T_{\text{flue,out}} - T_{\text{ref}})$$

The energy losses are given as 12% of the total energy, so the energy loss fraction, η, is 0.12. The total energy losses are

$$\dot{Q}_L = \eta \rho_{\text{flue,in}} \dot{V}_{\text{flue,in}} c_{p,\text{mean}} (T_{\text{flue,in}} - T_{\text{ref}})$$

The energy generated by the waste during combustion is

$$\dot{Q}_{c,\text{waste,in}} = \rho_{\text{waste,in}} \dot{V}_{\text{waste,in}} \Delta h_{c,\text{waste,in}}$$

The volumetric flow rate of the auxiliary fuel is $\dot{V}_{\text{aux}}$. The auxiliary fuel heat content is

$$\dot{Q}_{c,\text{aux}} = \rho_{\text{aux}} \dot{V}_{\text{aux}} \Delta h_{c,\text{aux}}$$

An energy balance around the entire incinerator gives

$$\begin{aligned} 0 \text{ Btu/min} &= \text{energy in} - \text{energy out} - \text{energy generated} \\ &= \dot{Q}_{\text{waste,in}} - (\dot{Q}_{\text{flue,out}} + \dot{Q}_L) \\ &\quad - (\dot{Q}_{c,\text{waste,in}} + \dot{Q}_{c,\text{aux}}) \end{aligned}$$

Substituting the five preceding equations into this energy balance equation gives

$$\begin{aligned} 0 \text{ Btu/min} = {} & \rho_{\text{waste,in}} \dot{V}_{\text{waste,in}} c_{p,\text{mean}} (T_{\text{waste,in}} - T_{\text{ref}}) \\ & - \rho_{\text{flue,out}} \dot{V}_{\text{flue,out}} c_{p,\text{mean}} (T_{\text{flue,out}} - T_{\text{ref}}) \\ & - \eta \rho_{\text{flue,in}} \dot{V}_{\text{flue,in}} c_{p,\text{mean}} (T_{\text{flue,in}} - T_{\text{ref}}) \\ & - \rho_{\text{waste,in}} \dot{V}_{\text{waste,in}} \Delta h_{c,\text{waste,in}} \\ & - \rho_{\text{aux}} \dot{V}_{\text{aux}} \Delta h_{c,\text{aux}} \end{aligned}$$

The mass flow of the waste gas is

$$\dot{m}_{\text{waste,in}} = \rho_{\text{waste,in}} Q_{\text{waste,in}}$$

The mass flow rate of the flue gas out of the preheater is

$$\dot{m}_{\text{flue,out}} = \rho_{\text{flue,out}} Q_{\text{flue,out}}$$

The mass flow rate of the auxiliary fuel is

$$\dot{m}_{\text{aux}} = \rho_{\text{aux}} Q_{\text{aux}}$$

The total mass balance around the entire incinerator is

$$\begin{aligned} \text{mass auxiliary fuel} + \text{mass waste feed} &= \text{mass flue gas} \\ \dot{m}_{\text{aux}} + \dot{m}_{\text{waste,in}} &= \dot{m}_{\text{flue,out}} \\ \rho_{\text{aux}} Q_{\text{aux}} + \rho_{\text{waste,in}} Q_{\text{waste,in}} &= \rho_{\text{flue,out}} Q_{\text{flue,out}} \end{aligned}$$

The preheater does not change the mass flow rate of the flue gas, so the mass flow rate of the flue gas into the preheater equals the mass flow rate of the flue gas out of the preheater. Therefore,

$$\rho_{\text{aux}} Q_{\text{aux}} + \rho_{\text{waste,in}} Q_{\text{waste,in}} = \rho_{\text{flue,in}} Q_{\text{flue,in}}$$

Substituting these values into the energy balance equation,

$$\begin{aligned} 0 \text{ Btu/lbm} = {} & \rho_{\text{waste,in}} Q_{\text{waste,in}} c_{p,\text{mean}} (T_{\text{waste,in}} - T_{\text{ref}}) \\ & - (\rho_{\text{aux}} Q_{\text{aux}} + \rho_{\text{waste,in}} Q_{\text{waste,in}}) \\ & \times c_{p,\text{mean}} (T_{\text{flue,out}} - T_{\text{ref}}) \\ & - \eta (\rho_{\text{aux}} Q_{\text{aux}} + \rho_{\text{waste,in}} Q_{\text{waste,in}}) \\ & \times c_{p,\text{mean}} (T_{\text{flue,in}} - T_{\text{ref}}) \\ & - \rho_{\text{waste,in}} Q_{\text{waste,in}} \Delta H_{c,\text{waste,in}} \\ & - \rho_{\text{aux}} Q_{\text{aux}} \Delta H_{c,\text{aux}} \end{aligned}$$

Rearranging and solving for the volumetric flow rate of auxiliary fuel gives

$$\dot{V}_{aux} = \frac{\rho_{waste,in}\dot{V}_{waste,in}\left(\begin{array}{l} c_{p,mean} \\ \times\left(\begin{array}{l} T_{waste,in} - T_{flue,out} \\ -\eta(T_{flue,in} - T_{ref}) \end{array}\right) \\ -\Delta h_{c,waste,in} \end{array}\right)}{\rho_{aux}\left(\begin{array}{l} c_{p,mean} \\ \times\left(\begin{array}{l} T_{flue,out} - T_{ref} \\ +\eta(T_{flue,in} - T_{ref}) \end{array}\right) \\ +\Delta h_{c,aux} \end{array}\right)}$$

$$= \frac{\left(0.0739\ \frac{lbm}{ft^3}\right)\left(25{,}000\ \frac{ft^3}{min}\right) \times\left(\begin{array}{l}\left(0.255\ \frac{Btu}{lbm\text{-}^\circ F}\right)\left(\begin{array}{l}110^\circ F - 482.5^\circ F - (0.12) \\ \times(1600^\circ F - 77^\circ F)\end{array}\right) \\ -\left(-58.47\ \frac{Btu}{lbm}\right)\end{array}\right)}{\left(0.0408\ \frac{lbm}{ft^3}\right) \times\left(\begin{array}{l}\left(0.255\ \frac{Btu}{lbm\text{-}^\circ F}\right)\left(\begin{array}{l}482.5^\circ F - 77^\circ F + (0.12) \\ \times(1600^\circ F - 77^\circ F)\end{array}\right) \\ +\left(-19{,}500\ \frac{Btu}{lbm}\right)\end{array}\right)}$$

$$= 194.516\ ft^3/min \quad (190\ ft^3/min)$$

The answer is (A).

4. The density of the flue gas, $\rho_{flue\ gas}$, is given as 1350 kg/m^3. The density of the water, ρ_{water}, is given as 1000 kg/m^3. From *NCEES Handbook:* Density, the specific gravity is

$$SG = \frac{\rho}{1000\ \frac{kg}{m^3}}$$

Applying the preceding equation to the conditions of the problem statement, the specific gravity of the flue gas is

$$SG = \frac{\rho_{flue\ gas}}{\rho_{water}} = \frac{1350\ \frac{kg}{m^3}}{1000\ \frac{kg}{m^3}} = 1.35$$

The fluid head, h, is given as 32 mm Hg, and the specific gravity of mercury, SG_{Hg}, is 13.6. The head loss through the venturi meter is

$$\Delta h_{venturi} = h\left(\frac{SG_{Hg}}{SG} - 1\right) = \frac{(32\ mm)\left(\frac{13.6}{1.35} - 1\right)}{1000\ \frac{mm}{m}} = 0.290\ m$$

The ratio of the diameter of the throat to the diameter of the pipe is

$$\beta = \frac{d_{throat}}{d_{pipe}} = \frac{75\ mm}{150\ mm} = 0.5$$

The coefficient of the venturi meter, C, is given as 0.985. The velocity of the fluid through the throat of the venturi meter is

$$u = C\sqrt{\frac{2g\Delta h_{venturi}}{1-\beta^4}} = 0.985\sqrt{\frac{(2)\left(9.81\ \frac{m}{s^2}\right)(0.290\ m)}{1-0.5^4}} = 2.427\ m/s$$

The cross-sectional area at the throat of the venturi meter is

$$A = \frac{\pi d_{throat}^2}{4} = \frac{\pi\left(\frac{75\ mm}{1000\ \frac{mm}{m}}\right)^2}{4} = 0.00442\ m^2$$

From *NCEES Handbook:* Conservation of Mass, the continuity equation is

$$\rho_1 A_1 u_1 = \rho_2 A_2 u_2$$

Applying the preceding equation to the conditions of the problem statement, the mass flow rate of the flue gas is

$$\dot{Q} = \rho_{flue\ gas}Au = \left(1350\ \frac{kg}{m^3}\right)(0.00442\ m^2)\left(2.427\ \frac{m}{s}\right) = 14.48\ kg/s \quad (14\ kg/s)$$

The answer is (B).

5. The inlet temperature is

$$T = 450°\text{F} + 460° = 910°\text{R}$$

The pressure at inlet conditions, P, is given as 5.2 atm. The universal gas constant, R, is 0.7302 atm-ft^3/lb mole-°R. Because the feed is equimolecular in M and Q, the mole fraction of M in the feed, y, is 0.50. From *NCEES Handbook:* Ideal Gas Law, the ideal gas law is

$$PV = nRT$$

Rearranging the preceding equation gives the concentration of a pure gas.

$$\frac{n}{V} = \frac{P}{RT} = C$$

Applying the preceding equation to a mixture of M and N, the initial concentration of M is

$$\begin{aligned} C_{\text{M},0} &= y\left(\frac{P}{RT}\right) \\ &= (0.50)\left(\frac{5.2 \text{ atm}}{\left(0.7302 \ \frac{\text{atm-ft}^3}{\text{lb mole-°R}}\right)(910°\text{R})}\right) \\ &= 0.00391 \text{ lb mole/ft}^3 \end{aligned}$$

The molar feed rate, F, is 15 lb mole/sec. Because the feed is equimolecular in M and Q, the molar feed rate of M is half the total feed rate.

$$F_{\text{M},0} = \frac{F}{2} = \frac{15 \ \frac{\text{lb mole}}{\text{sec}}}{2} = 7.5 \text{ lb mole/sec}$$

The reaction rate constant for the reaction M → N is

$$k = \frac{1200 \text{ min}^{-1}}{60 \ \frac{\text{sec}}{\text{min}}} = 20 \text{ sec}^{-1}$$

The stoichiometric coefficients are 1 for products N, S, and U and −1 for reactants M and Q. The difference in stoichiometric coefficients is

$$\delta = 1 + 1 + 1 + (-1) + (-1) = 1$$

The fractional volume change is, by definition,

$$\varepsilon_\text{M} = y\delta = (0.50)(1) = 0.50$$

The desired conversion of M, X, is given as 0.60.

From *NCEES Handbook:* First-Order Reactions, the following relationship applies.

$$k\tau = -(1 + \varepsilon_\text{A})\ln(1 - X_\text{A}) - \varepsilon_\text{A} X_\text{A}$$

Rearranging,

$$\tau = \left(\frac{1}{k}\right)(1 + \varepsilon_\text{A})\ln\frac{1}{1 - X_\text{A}} - \varepsilon_\text{A} X_\text{A}$$

From *NCEES Handbook:* Plug–Flow Reactor, the following relationship applies.

$$\tau = \frac{C_{\text{Ao}} V_{\text{PFR}}}{F_{\text{Ao}}}$$

Combining the two preceding equations gives

$$\frac{C_{\text{Ao}} V_{\text{PFR}}}{F_{\text{Ao}}} = \left(\frac{1}{k}\right)(1 + \varepsilon_\text{A})\ln\frac{1}{1 - X_\text{A}} - \varepsilon_\text{A} X_\text{A}$$

Solving for the PFR volume gives

$$\begin{aligned} V_{\text{PFR}} &= \left(\frac{F_{\text{Ao}}}{kC_{\text{Ao}}}\right)(1 + \varepsilon_\text{A})\ln\frac{1}{1 - X_\text{A}} - \varepsilon_\text{A} X_\text{A} \\ &= \left(\frac{F_{\text{M},0}}{kC_{\text{M},0}}\right)\left((1 + \varepsilon_\text{M})\ln\frac{1}{1 - X_\text{M}} - \varepsilon_\text{M} X_\text{M}\right) \\ &= \left(\frac{7.5 \ \frac{\text{lb mole}}{\text{sec}}}{(20 \text{ sec}^{-1})\left(0.00391 \ \frac{\text{lb mole}}{\text{ft}^3}\right)}\right) \\ &\quad \times \left((1 + 0.50)\left(\ln\frac{1}{1 - 0.60}\right) - (0.50)(0.60)\right) \\ &= 103.0 \text{ ft}^3 \quad (100 \text{ ft}^3) \end{aligned}$$

The answer is (B).

6. From *NCEES Handbook:* Properties of Water, the molar mass of water is 18.01528 lbm/lb mole.

From *NCEES Handbook:* Material Balances With Reaction, the molecular weight of dry air is 28.965 lbm/lb mole. The average heat capacity of the dry air, c_p, is 0.24 Btu/lbm-°F.

The total pressure, P, is given as 14.7 lbf/in^2. From *NCEES Handbook* table "Saturated Steam (U.S. Units)—Temperature Table," the partial pressure of the water vapor in the saturated air at 50°F, P_{wv}, is 0.17796 lbf/in^2. The dry air, the water vapor, and the mixture are to be treated as ideal gases, so the specific humidity of the saturated air at 50°F is

$$\omega_{out} = \left(\frac{P_{wv}}{P_{dry\,air}}\right)\left(\frac{MW_{H_2O}}{MW_{air}}\right)$$

The pressure of the ambient air entering the saturator, P, is the same as the pressure of the saturated air leaving the saturator. Dalton's law applies, so the total pressure of the saturated air is

$$P = P_{dry\,air} + P_{wv}$$

Therefore,

$$\begin{aligned} P_{dry\,air} &= P - P_{wv} \\ \omega_{out} &= \left(\frac{P_{wv}}{P - P_{wv}}\right)\left(\frac{MW_{H_2O}}{MW_{air}}\right) \\ &= \left(\frac{0.17796\ \frac{\text{lbf}}{\text{in}^2}}{14.7\ \frac{\text{lbf}}{\text{in}^2} - 0.17796\ \frac{\text{lbf}}{\text{in}^2}}\right)\left(\frac{18.01528\ \frac{\text{lbm water}}{\text{lb mole}}}{28.965\ \frac{\text{lbm dry air}}{\text{lb mole}}}\right) \\ &= 0.00762\ \text{lbm water/lbm dry air} \end{aligned}$$

From *NCEES Handbook:* "Saturated Steam (U.S. Units)—Temperature Table," the specific enthalpy of the 50°F water vapor, h_{out}, is 1083.4 Btu/lbm, the specific enthalpy of the 50°F makeup water, h_{feed}, is 18.054 Btu/lbm. and the specific enthalpy of the 75°F water vapor at the inlet, h_{in}, is 1094.3 Btu/lbm. The specific humidity at inlet conditions is ω_{in}. The ambient air changes the enthalpy in the saturator due to temperature change and due to water incorporation. As the air enters the adiabatic saturator at a temperature T_{in} and leaves the saturator at T_{out}, the change of specific enthalpy of the air due to temperature change alone is $c_p(T_{in} - T_{out})$. Because the air incorporates water in the evaporator, the specific enthalpy change of the air due to the incorporation of water alone is $\omega_{out}(h_{out} - h_{feed})$. The saturated air specific enthalpy change at the exit is $\omega_{in}(h_{in} - h_{feed})$. An energy balance around the adiabatic saturator gives

$$c_p(T_{in} - T_{out}) + \omega_{out}(h_{out} - h_{feed}) = \omega_{in}(h_{in} - h_{feed})$$

At the inlet temperature, the specific humidity is

$$\begin{aligned} \omega_{in} &= \frac{c_p(T_{in} - T_{out}) + \omega_{out}(h_{out} - h_{feed})}{h_{in} - h_{feed}} \\ &= \frac{\left(\begin{array}{l}\left(0.24\ \frac{\text{Btu}}{\text{lbm-°F}}\right)(75\text{°F} - 50\text{°F}) \\ +\left(0.00762\ \frac{\text{lbm water}}{\text{lbm dry air}}\right) \\ \times\left(1083.4\ \frac{\text{Btu}}{\text{lbm}} - 18.054\ \frac{\text{Btu}}{\text{lbm}}\right)\end{array}\right)}{1094.3\ \frac{\text{Btu}}{\text{lbm}} - 18.054\ \frac{\text{Btu}}{\text{lbm}}} \\ &= 0.01312\ \text{lbm water/lbm dry air} \end{aligned}$$

The partial pressure of water vapor at the inlet temperature is

$$\begin{aligned} P_{wv,in} &= \frac{\omega_{in}P}{\omega_{in} + \frac{MW_{H_2O}}{MW_{air}}} \\ &= \frac{\left(0.01312\ \frac{\text{lbm water}}{\text{lbm dry air}}\right)\left(14.7\ \frac{\text{lbf}}{\text{in}^2}\right)}{0.01312\ \frac{\text{lbm water}}{\text{lbm dry air}} + \frac{18.01528\ \frac{\text{lbm water}}{\text{lb mole}}}{28.965\ \frac{\text{lbm dry air}}{\text{lb mole}}}} \\ &= 0.3037\ \text{lbf/in}^2 \end{aligned}$$

From *NCEES Handbook:* "Saturated Steam (U.S. Units)—Temperature Table," the saturation pressure of water vapor at 75°F, $P_{sat,in}$, is 0.42964 lbf/in^2. The relative humidity of the ambient air is

$$\begin{aligned} \phi &= \frac{P_{wv,in}}{P_{sat,in}} = \frac{0.3037\ \frac{\text{lbf}}{\text{in}^2}}{0.42964\ \frac{\text{lbf}}{\text{in}^2}} \\ &= 0.707 \quad (0.71) \end{aligned}$$

The answer is (B).

7. The density of water, ρ_{water}, is 62.4 lbm/ft^3. The pressure at the pitot tube, the pressure at the manometer, and the barometric pressure are given as

$$\begin{aligned} P_{pitot} &= 0.70\ \text{in wg} \\ P_{man} &= -0.60\ \text{in wg} \\ P_{bar} &= 725\ \text{mm Hg} \end{aligned}$$

The pressure drop in the smokestack is

$$\begin{aligned}\Delta P &= (P_{\text{pitot}} - P_{\text{man}})\rho_{\text{water}}\left(\frac{g}{g_c}\right)\\ &= \left(\frac{0.70\text{ in} - (-0.60\text{ in})}{12\ \dfrac{\text{in}}{\text{ft}}}\right)\left(62.4\ \frac{\text{lbm}}{\text{ft}^3}\right)\\ &\quad\times\left(\frac{32.2\ \dfrac{\text{ft}}{\text{sec}^2}}{32.2\ \dfrac{\text{ft-lbm}}{\text{lbf-sec}^2}}\right)\\ &= 6.76\text{ lbf/ft}^2\end{aligned}$$

The temperature of the waste gas stream is given as

$$T_{\text{gas}} = 270^\circ\text{F} + 460^\circ = 730^\circ\text{R}$$

The molecular weight of the gas stream, MW_{gas}, is 29 lbm/lb mole. The volume of any ideal gas at standard conditions, V, is 359 ft^3. The number of moles of ideal gas, n, is 1 lb mole. The standard temperature, T_{std}, is 492°R, and the standard pressure, P_{std}, is 760 mm Hg. From *NCEES Handbook:* Density and Relative Density, the density is

$$\rho = \frac{m}{V}$$

From *NCEES Handbook:* Ideal Gas law, for an ideal gas

$$PV = \frac{mRT}{MW}$$

Combining the preceding equations gives

$$\frac{m}{V} = \left(\frac{n}{V}\right)(MW_{\text{gas}})\left(\frac{T_{\text{std}}}{T_{\text{gas}}}\right)\left(\frac{P_{\text{bar}}}{P_{\text{std}}}\right)$$

Using the ideal gas law, the density of the waste gas stream is

$$\begin{aligned}\rho_{\text{gas}} &= \left(\frac{n}{V}\right)(MW_{\text{gas}})\left(\frac{T_{\text{std}}}{T_{\text{gas}}}\right)\left(\frac{P_{\text{bar}}}{P_{\text{std}}}\right)\\ &= \left(\frac{1\text{ lb mole}}{359\text{ ft}^3}\right)\left(29\ \frac{\text{lbm}}{\text{lb mole}}\right)\left(\frac{492^\circ\text{R}}{730^\circ\text{R}}\right)\left(\frac{725\text{ mm Hg}}{760\text{ mm Hg}}\right)\\ &= 0.0519\text{ lbm/ft}^3\end{aligned}$$

As in *NCEES Handbook:* Orifice Discharging Freely into Atmosphere, the maximum velocity of the waste gas stream is

$$\begin{aligned}u_{\max} &= \sqrt{2gh} = \sqrt{2g\left(\frac{\Delta P}{\rho_{\text{gas}}}\right)\left(\frac{g_c}{g}\right)} = \sqrt{\frac{2g_c\Delta P}{\rho_{\text{gas}}}}\\ &= \sqrt{\frac{(2)\left(32.2\ \dfrac{\text{ft-lbm}}{\text{lbf-sec}^2}\right)\left(6.76\ \dfrac{\text{lbf}}{\text{ft}^2}\right)}{0.0519\ \dfrac{\text{lbm}}{\text{ft}^3}}}\\ &= 91.6\text{ ft/sec}\end{aligned}$$

The average velocity of the waste gas stream is 81% of its maximum velocity.

$$\begin{aligned}u_{\text{avg}} &= 0.81u_{\max}\\ &= (0.81)\left(91.6\ \frac{\text{ft}}{\text{sec}}\right)\\ &= 74.2\text{ ft/sec}\end{aligned}$$

The area of the smokestack is

$$A = \frac{\pi D^2}{4} = \frac{\pi(5\text{ ft})^2}{4} = 19.6\text{ ft}^2$$

The waste gas volumetric flow rate is

$$\begin{aligned}\dot{Q} &= u_{\text{avg}}A\\ &= \left(74.2\ \frac{\text{ft}}{\text{sec}}\right)(19.6\text{ ft}^2)\\ &= 1454\text{ ft}^3/\text{sec}\end{aligned}$$

The fraction of sulfur dioxide in the waste gas stream is

$$f_{SO_2} = \frac{0.6\%}{100\%} = 0.006$$

The molecular weight of sulfur dioxide, MW_{SO_2}, is 64 lbm/lb mole. The mass flow rate of SO_2 in the smokestack is

$$\begin{aligned}\dot{m}_{SO_2} &= \dot{Q}\rho_{gas}\left(\frac{MW_{SO_2}}{MW_{gas}}\right)f_{SO_2} \\ &= \left(\left(1454\ \frac{ft^3}{sec}\right)\left(3600\ \frac{sec}{hr}\right)\right) \\ &\quad \times\left(0.0519\ \frac{lbm}{ft^3}\right)\left(\frac{64\ \frac{lbm}{lb\ mole}}{29\ \frac{lbm}{lb\ mole}}\right)(0.006) \\ &= 3597\ lbm/hr \quad (3600\ lbm/hr)\end{aligned}$$

The answer is (C).

8. The balanced reaction is

$$C_2H_2 + N_2 \rightarrow 2HCN$$

The rate of production of HCN is given as

$$\dot{m}_{HCN} = 1000\ lbm/hr$$

From the stoichiometry of the reaction, 2 lb mole of HCN are produced per 1 lb mole of C_2H_2 in the feed. The mass rate of C_2H_2 needed to react to produce 1000 lbm/hr of HCN is

$$\begin{aligned}\dot{m}_{C_2H_2} &= \frac{\dot{m}_{HCN}(MW_{C_2H_2})}{2(MW_{HCN})} \\ &= \frac{\left(1000\ \frac{lbm}{hr}\right)\left(26\ \frac{lbm}{lb\ mole}\right)}{(2)\left(27\ \frac{lbm}{lb\ mole}\right)} \\ &= 481.48\ lbm/hr\end{aligned}$$

From the stoichiometry of the reaction, 1 lb mole of N_2 reacts with 1 lb mole of C_2H_2. The minimum mass rate of N_2 required to consume all of the C_2H_2 in the feed is

$$\begin{aligned}\dot{m}_{N_2,reacted} &= \frac{\dot{m}_{C_2H_2}(MW_{N_2})}{MW_{C_2H_2}} \\ &= \frac{\left(481.48\ \frac{lbm}{hr}\right)\left(28\ \frac{lbm}{lb\ mole}\right)}{26\ \frac{lbm}{lb\ mole}} \\ &= 518.52\ lbm/hr\end{aligned}$$

The maximum capacity of the recycled steam is 52 lbm/hr, so the mass rate of N_2 in the recycled stream is

$$\dot{m}_{N_2,recycled} = 52\ lbm/hr$$

From *NCEES Handbook:* Material Balances, the material balance with no reaction is

Accumulation = Input − Output + Generation − Consumption

An N_2 mass balance around the mixer gives the mass rate of N_2 in the feed.

$$\begin{aligned}\dot{m}_{N_2,feed} &= \dot{m}_{N_2,reacted} + \dot{m}_{N_2,recycled} \\ &= 518.52\ \frac{lbm}{hr} + 52\ \frac{lbm}{hr} \\ &= 570.52\ lbm/hr\end{aligned}$$

The percentage of excess N_2 in the feed is

$$\begin{aligned}P &= \frac{\dot{m}_{N_2,feed} - \dot{m}_{N_2,reacted}}{\dot{m}_{N_2,feed}} \times 100\% \\ &= \frac{570.52\ \frac{lbm}{hr} - 518.52\ \frac{lbm}{hr}}{570.52\ \frac{lbm}{hr}} \times 100\% \\ &= 9.11\% \quad (9.1\%)\end{aligned}$$

The answer is (B).

9. The relevant data are shown in the illustration.

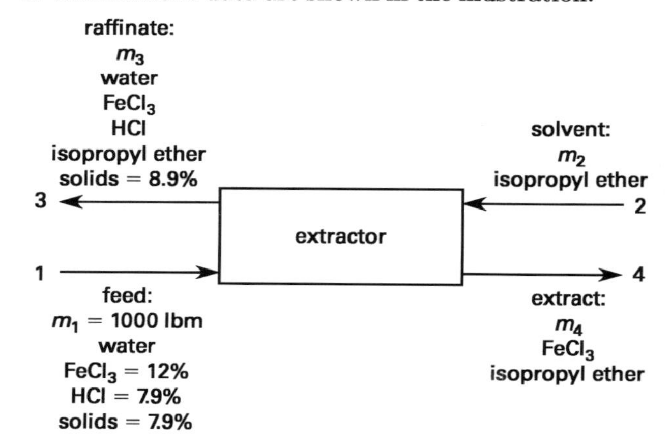

Throughout the solution the *NCEES Handbook:* Material Balances, the balance equation is used.

Accumulation = Input − Output + Generation − Consumption

The feed contains 7.9% solids, the raffinate contains 8.9% solids, and there are no solids in the extract.

Therefore, a mass balance of solids around the extractor gives

$$
\begin{aligned}
0.079m_1 &= 0.089m_3 \\
m_3 &= \frac{0.079m_1}{0.089} \\
&= \frac{(0.079)(1000\ \text{lbm})}{0.089} \\
&= 887.64\ \text{lbm}
\end{aligned}
$$

Hydrochloric acid (HCl) is completely soluble in water, so the raffinate contains all the HCl from the feed. The feed is 7.9% HCl, so a mass balance of HCl around the extractor gives

$$m_{1,\text{HCl}} = 0.079F_1 = (0.079)(1000\ \text{lbm}) = 79\ \text{lbm}$$

The feed is 12% (by weight) ferric chloride ($FeCl_3$), so the mass of $FeCl_3$ in the feed is

$$m_{1,\text{FeCL}_3} = 0.12F_1 = (0.12)(1000\ \text{lbm}) = 120\ \text{lbm}$$

The feed is 7.9% (by weight) inert solids, so the mass of inert solids in the feed is

$$m_{1,\text{solids}} = 0.079m_1 = (0.079)(1000\ \text{lbm}) = 79\ \text{lbm}$$

The rest of the feed is water, so the mass of water in the feed is

$$
\begin{aligned}
m_{1,\text{water}} &= m_1 - m_{1,\text{HCl}} - m_{1,\text{FeCl}_3} - m_{1,\text{solids}} \\
&= 1000\ \text{lbm} - 79\ \text{lbm} - 120\ \text{lbm} - 79\ \text{lbm} \\
&= 722\ \text{lbm}
\end{aligned}
$$

The extract (stream 4) does not contain water, so all the water that enters in the feed leaves in the raffinate (stream 3). A mass balance of water around the extractor gives

$$m_{3,\text{water}} = m_{1,\text{water}} = 722\ \text{lbm}$$

The extract does not contain solids, so all the solids that enter in the feed leave in the raffinate. A mass balance of solids around the extractor gives

$$m_{3,\text{solids}} = m_{1,\text{solids}} = 79\ \text{lbm}$$

The desired recovery of $FeCl_3$ is 99%, so the minimum mass of $FeCl_3$ in the extract is

$$
\begin{aligned}
m_{4,\text{FeCl}_3} &= 0.99m_{1,\text{FeCl}_3} \\
&= (0.99)(120\ \text{lbm}) \\
&= 118.8\ \text{lbm}
\end{aligned}
$$

The maximum mass of $FeCl_3$ in the raffinate is

$$
\begin{aligned}
m_{3,\text{FeCl}_3} &= m_{1,\text{FeCl}_3} - m_{4,\text{FeCl}_3} \\
&= 120\ \text{lbm} - 118.8\ \text{lbm} \\
&= 1.2\ \text{lbm}
\end{aligned}
$$

The extract does not contain HCl, so all the HCl that enters in the feed leaves in the raffinate.

$$m_{3,\text{HCl}} = m_{1,\text{HCl}} = 79\ \text{lbm}$$

The masses of the other components in the raffinate have now been found, so the mass of the remaining component, isopropyl ether, is

$$
\begin{aligned}
m_{3,\text{ether}} &= m_3 - m_{3,\text{water}} - m_{3,\text{solids}} - m_{3,\text{HCl}} - m_{3,\text{FeCl}_3} \\
&= 887.64\ \text{lbm} - 722\ \text{lbm} - 79\ \text{lbm} \\
&\quad -79\ \text{lbm} - 1.2\ \text{lbm} \\
&= 6.44\ \text{lbm}
\end{aligned}
$$

The ratio of the mass of $FeCl_3$ to the mass of isopropyl ether will be the same in both the raffinate (stream 3) and the extract (stream 4), so

$$
\begin{aligned}
\frac{m_{3,\text{FeCl}_3}}{m_{3,\text{ether}}} &= \frac{m_{4,\text{FeCl}_3}}{m_{4,\text{ether}}} \\
m_{4,\text{ether}} &= \frac{m_{3,\text{ether}} m_{4,\text{FeCl}_3}}{m_{3,\text{FeCl}_3}} \\
&= \frac{(6.44\ \text{lbm})(118.8\ \text{lbm})}{1.2\ \text{lbm}} \\
&= 637.56\ \text{lbm}
\end{aligned}
$$

A mass balance of the isopropyl ether around the extractor gives

$$
\begin{aligned}
m_2 &= m_{3,\text{ether}} + m_{4,\text{ether}} \\
&= 6.44\ \text{lbm} + 637.56\ \text{lbm} \\
&= 644\ \text{lbm} \quad (640\ \text{lbm})
\end{aligned}
$$

The answer is (C).

10. Throughout this solution the balance equation from *NCEES Handbook:* Material Balances is used repeatedly.

$$\textit{Accumulation} = \textit{Input} - \textit{Output} + \textit{Generation} - \textit{Consumption}$$

The relevant data are shown in the illustration.

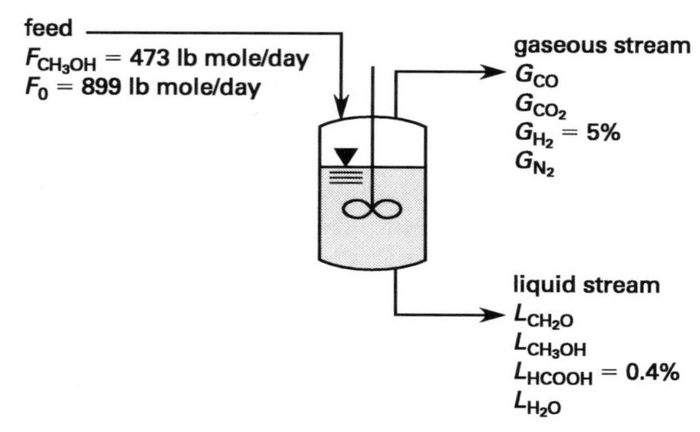

The air is 21% O_2, so the molar flow rate of O_2 in the feed is

$$F_{O_2} = 0.21F_{air} = (0.21)\left(899\ \frac{\text{lb mole}}{\text{day}}\right) = 189 \text{ lb mole/day}$$

The air is 79% N_2, so the molar flow rate of N_2 in the feed is

$$F_{N_2} = 0.79F_{air} = (0.79)\left(899\ \frac{\text{lb mole}}{\text{day}}\right) = 710 \text{ lb mole/day}$$

Nitrogen is inert, and none of it is in the liquid stream, so the molar flow rates of N_2 in the feed stream and the gaseous stream must be equal.

$$G_{N_2} = F_{N_2} = 710 \text{ lb mole/day}$$

The H_2 in the gaseous stream is 5% of the total gaseous stream, so the molar flow rate of H_2 in the gaseous stream is

$$G_{H_2} = 0.05G$$

The gaseous stream, then, contains

$$\begin{aligned} G &= G_{CO} + G_{CO_2} + G_{H_2} + G_{N_2} \\ &= G_{CO} + G_{CO_2} + 0.05G + 710\ \frac{\text{lb mole}}{\text{day}} \end{aligned}$$

Subtracting G from each side,

$$\begin{aligned} 0\ \frac{\text{lb mole}}{\text{day}} &= G_{CO} + G_{CO_2} - 0.95G + 710\ \frac{\text{lb mole}}{\text{day}} \\ -710\ \frac{\text{lb mole}}{\text{day}} &= G_{CO} + G_{CO_2} - 0.95G \end{aligned}$$

The conversion of CH_3OH is 60%, so the molar flow rate of CH_3OH in the liquid stream is

$$\begin{aligned} L_{CH_3OH} &= (1 - 0.60)F_{CH_3OH} \\ &= (1 - 0.60)\left(473\ \frac{\text{lb mole}}{\text{day}}\right) \\ &= 189 \text{ lb mole/day} \end{aligned}$$

The molar flow rates of CH_3OH and CH_2O in the liquid stream are equal, so

$$L_{CH_2O} = L_{CH_3OH} = 189 \text{ lb mole/day}$$

The HCOOH in the liquid stream is 0.4% of the total liquid stream, so the molar flow rate of HCOOH in the liquid stream is

$$L_{HCOOH} = 0.004L$$

The liquid stream, then, contains

$$\begin{aligned} L &= L_{CH_2O} + L_{CH_3OH} + L_{HCOOH} + L_{H_2O} \\ &= 189\ \frac{\text{lb mole}}{\text{day}} + 189\ \frac{\text{lb mole}}{\text{day}} + 0.004L + L_{H_2O} \end{aligned}$$

Subtracting L from each side,

$$\begin{aligned} 0\ \frac{\text{lb mole}}{\text{day}} &= 189\ \frac{\text{lb mole}}{\text{day}} + 189\ \frac{\text{lb mole}}{\text{day}} \\ &\quad - 0.996L + L_{H_2O} \\ -378\ \frac{\text{lb mole}}{\text{day}} &= L_{H_2O} - 0.996L \end{aligned}$$

The oxygen atom balance around the reactor gives

$$\begin{aligned} \text{moles of O in feed} &= \text{moles of O in gas} \\ &\quad + \text{moles of O in liquid} \\ 2F_{O_2} + F_{CH_3OH} &= G_{CO} + 2G_{CO_2} + L_{CH_2O} \\ &\quad + L_{CH_3OH} + 2L_{HCOOH} \\ &\quad + L_{H_2O} \end{aligned}$$

$$\begin{aligned} (2)\left(189\ \frac{\text{lb mole}}{\text{day}}\right) + 473\ \frac{\text{lb mole}}{\text{day}} &= G_{CO} + 2G_{CO_2} + 189\ \frac{\text{lb mole}}{\text{day}} \\ &\quad + 189\ \frac{\text{lb mole}}{\text{day}} + (2)(0.004L) \\ &\quad + L_{H_2O} \\ 473\ \frac{\text{lb mole}}{\text{day}} &= G_{CO} + 2G_{CO_2} + 0.008L + L_{H_2O} \end{aligned}$$

The hydrogen atom balance around the reactor gives

$$\begin{aligned}\text{moles of H in feed} &= \text{moles of H in gas} \\ &\quad +\text{moles of H in liquid} \\ 4F_{CH_3OH} &= 2G_{H_2} + 2L_{CH_2O} + 4L_{CH_3OH} \\ &\quad +2L_{HCOOH} + 2L_{H_2O} \\ (4)\left(473\ \frac{\text{lb mole}}{\text{day}}\right) &= (2)(0.05G) + (2)\left(189\ \frac{\text{lb mole}}{\text{day}}\right) \\ &\quad +(4)\left(189\ \frac{\text{lb mole}}{\text{day}}\right) \\ &\quad +(2)(0.004L) + 2L_{H_2O} \\ 758\ \frac{\text{lb mole}}{\text{day}} &= 0.1G + 0.008L + 2L_{H_2O}\end{aligned}$$

The carbon balance around the reactor gives

$$\begin{aligned}\text{moles of C in feed} &= \text{moles of C in gas} \\ &\quad +\text{moles of C in liquid} \\ F_{CH_3OH} &= G_{CO} + G_{CO_2} + L_{CH_2O} \\ &\quad +L_{CH_3OH} + L_{HCOOH} \\ 473\ \frac{\text{lb mole}}{\text{day}} &= G_{CO} + G_{CO_2} + 189\ \frac{\text{lb mole}}{\text{day}} \\ &\quad +189\ \frac{\text{lb mole}}{\text{day}} + 0.004L \\ 95\ \frac{\text{lb mole}}{\text{day}} &= G_{CO} + G_{CO_2} + 0.004L\end{aligned}$$

Take the following five equations found in the preceding solution and solve simultaneously.

$$\begin{aligned}G_{CO} + G_{CO_2} - 0.95G &= -710\ \frac{\text{lb mole}}{\text{day}} \\ L_{H_2O} - 0.996L &= -378\ \frac{\text{lb mole}}{\text{day}} \\ G_{CO} + 2G_{CO_2} + 0.008L + L_{H_2O} &= 473\ \frac{\text{lb mole}}{\text{day}} \\ 0.1G + 0.008L + 2L_{H_2O} &= 758\ \frac{\text{lb mole}}{\text{day}} \\ G_{CO} + G_{CO_2} + 0.004L &= 95\ \frac{\text{lb mole}}{\text{day}}\end{aligned}$$

The molar flow rate of water in the liquid stream is found to be

$$L_{H_2O} = 333.9\ \text{lb mole/day} \quad (330\ \text{lb mole/day})$$

The answer is (B).

11. The temperature difference between the inner and outer surfaces is

$$T_1 - T_2 = 1450\text{K} - 1024\text{K} = 426\text{K}$$

The area of the furnace wall is

$$A = hw = (0.48\ \text{m})(3.2\ \text{m}) = 1.536\ \text{m}^2$$

From *NCEES Handbook:* Fourier's Law of Conduction, the rate of heat transfer through the wall is given by the Fourier equation.

$$\begin{aligned}\dot{Q} &= \frac{kA}{\delta}(T_1 - T_2) = \frac{\left(2.20\ \frac{\text{W}}{\text{m}\cdot\text{K}}\right)(1.536\ \text{m}^2)(426\text{K})}{0.20\ \text{m}} \\ &= 7197.7\ \text{W} \quad (7200\ \text{W})\end{aligned}$$

The answer is (C).

12. A primary element is a synonym for a sensor.

The answer is <u>sensor</u>.

13. The total pressure, P, is given as 1200 mm Hg. The vapor pressure of pure pentane, p_5, is given as 1647.7 mm Hg. The vapor pressure of pure hexane, p_6, is given as 785.9 mm Hg. From *NCEES Handbook:* Distribution of Components Between Phases in a Vapor/ Liquid Equilibrium the distribution coefficient is

$$K_i = \frac{p_i^{\text{sat}}}{P}$$

Because Dalton's law and Raoult's law apply, the ratio of the mole fraction of pentane in the vapor phase to the mole fraction of pentane in the liquid phase is

$$K_5 = \frac{p_5^{\text{sat}}}{P} = \frac{1647.7\ \text{mm Hg}}{1200\ \text{mm Hg}} = 1.373$$

Because Dalton's law and Raoult's law apply, the ratio of the mole fraction of hexane in the vapor phase to the mole fraction of hexane in the liquid phase is

$$K_6 = \frac{p_6^{\text{sat}}}{P} = \frac{785.9\ \text{mm Hg}}{1200\ \text{mm Hg}} = 0.6549$$

The mole fraction of pentane in the liquid phase of the mixture is x. From *NCEES Handbook:* Bubble Point, for a mixture at its bubble point the following relationship applies.

$$\sum_{i=1}^{n} K_i x_i = 1$$

$$K_5 x + K_6(1 - x) = 1$$

$$x = \frac{1 - K_6}{K_5 - K_6} = \frac{1 - 0.6549}{1.373 - 0.6549} = 0.4806 \quad (0.48)$$

The answer is (A).

14. A combination of two or more instruments or control functions arranged so that signals pass from one to another for the purpose of measurement and/or control of a process variable is a loop.

The answer is <u>loop</u>.

15. The natural logarithms of the time and concentration data are

ln t	ln C_M
0.5306	6.6644
1.0986	5.5530
1.6677	4.4427
2.2192	3.3322
2.7726	2.2192
3.3322	1.0986

A plot of the natural logarithm of the time versus the natural logarithm of the concentration is shown.

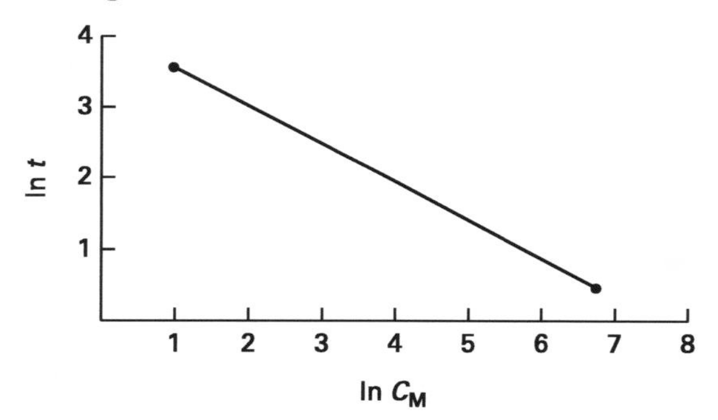

By least-squares analysis, the slope of the plot, m, is -0.5. The slope may also be calculated from a pair of data points. From *NCEES Handbook:* Linear Algebra, the point slope is

$$y - y_1 = m(x - x_1)$$

Solving for the slope gives

$$m = \frac{y - y_1}{x - x_1}$$

For instance, the second and third data points yield

$$m = \frac{1.6677 - 1.0986}{4.4427 - 5.5530} = -0.5$$

The order of the reaction is

$$n = 1 - m = 1 - (-0.5) = 1.5$$

The answer is (B).

16. The relevant data are shown in the illustration.

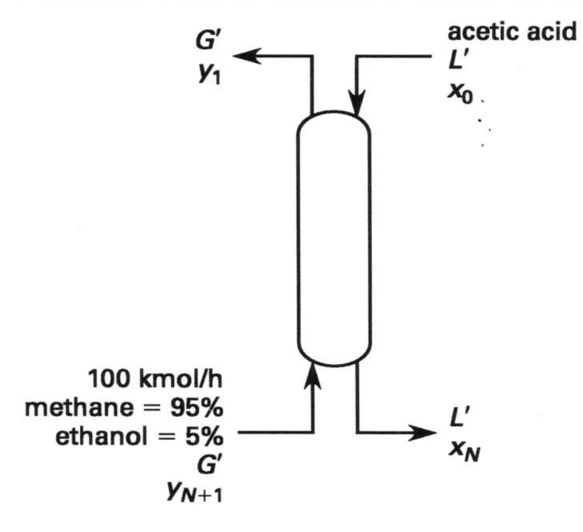

The liquid-phase activity coefficient of ethanol, γ, is given as 6. The vapor pressure of ethanol, p_v^{sat}, is given as 10.5 kPa. The total pressure, P, is given as 110 kPa. From *NCEES Handbook:* Distribution of Components Between Phases in a Vapor/Liquid Equilibrium, the distribution coefficient is defined as

$$K_i = \frac{p_i^{\text{sat}}}{P}$$

Considering the activity coefficient, the vapor-liquid equilibrium ratio is

$$K_N = \gamma\left(\frac{p_v^{\text{sat}}}{P}\right) = (6)\left(\frac{10.5 \text{ kPa}}{110 \text{ kPa}}\right) = 0.573$$

The molar flow rate of the entering gas, G_{in}, is 100 kmol/h. The inlet gas contains 95% methane, so the mole fraction of methane in the inlet gas, y_N, is 0.95. The molar flow rate of solute-free gas (carrier gas) is

$$\begin{aligned} G' &= y_N G_{\text{in}} \\ &= (0.95)\left(100\ \frac{\text{kmol}}{\text{h}}\right) \\ &= 95\ \text{kmol/h} \end{aligned}$$

The required recovery of the ethanol is given as 90%, so the mole fraction of ethanol recovered, r, is 0.90. The minimum solute-free absorbent rate is

$$\begin{aligned} L'_{\text{min}} &= G'K_N r \\ &= \left(95\ \frac{\text{kmol}}{\text{h}}\right)(0.573)(0.90) \\ &= 48.992\ \text{kmol/h} \end{aligned}$$

The tower is operating at an actual liquid molar flow rate 1.5 times the minimum, so the actual liquid molar flow rate is

$$\begin{aligned} L' &= 1.5L'_{\text{min}} \\ &= (1.5)\left(48.992\ \frac{\text{kmol}}{\text{h}}\right) \\ &= 73.488\ \text{kmol/h} \end{aligned}$$

The entering feed is 5% ethanol, so the molar flow rate of ethanol in the feed is

$$\begin{aligned} n_{\text{in}} &= 0.05G_{\text{in}} \\ &= (0.05)\left(100\ \frac{\text{kmol}}{\text{h}}\right) \\ &= 5\ \text{kmol/h} \end{aligned}$$

The amount of ethanol transferred from the gas to the liquid is 90% of the amount of ethanol in the entering gas. So, the molar flow rate of ethanol transferred is

$$\begin{aligned} n_{\text{out}} &= 0.90n_{\text{in}} \\ &= (0.90)\left(5\ \frac{\text{kmol}}{\text{h}}\right) \\ &= 4.5\ \text{kmol/h} \end{aligned}$$

From *NCEES Handbook:* Material Balances, the general balance equation is

$$\textit{Accumulation} = \textit{Input} - \textit{Output} + \textit{Generation} - \textit{Consumption}$$

This equation is used extensively throughout this problem solution.

An ethanol mass balance around the entire column gives the molar flow rate of ethanol remaining in the exiting gas.

$$\begin{aligned} n_{\text{remaining}} &= n_{\text{in}} - n_{\text{out}} \\ &= 5\ \frac{\text{kmol}}{\text{h}} - 4.5\ \frac{\text{kmol}}{\text{h}} \\ &= 0.5\ \text{kmol/h} \end{aligned}$$

The acetic acid contains no ethanol, so the mole fraction of ethanol in the liquid phase entering the column, x_0, is zero. The mole fraction of ethanol leaving the column is

$$\begin{aligned} y_1 &= \frac{n_{\text{remaining}}}{G'} \\ &= \frac{0.5\ \frac{\text{kmol}}{\text{h}}}{95\ \frac{\text{kmol}}{\text{h}}} \\ &= 0.00526 \end{aligned}$$

The overall mass balance around the tower gives

$$x_0L' + y_{N+1}G' = x_NL' + y_1G'$$

Replacing,

$$\begin{aligned} (0)\left(73.488\ \frac{\text{kmol}}{\text{h}}\right) + y_{N+1}\left(95\ \frac{\text{kmol}}{\text{h}}\right) \\ = x_N\left(73.488\ \frac{\text{kmol}}{\text{h}}\right) + (0.00526)\left(95\ \frac{\text{kmol}}{\text{h}}\right) \end{aligned}$$

Simplifying and rearranging gives the equation of the operating line.

$$\begin{aligned} y_{N+1} &= x_N\left(\frac{73.488\ \frac{\text{kmol}}{\text{h}}}{95\ \frac{\text{kmol}}{\text{h}}}\right) + (0.00526)\left(\frac{95\ \frac{\text{kmol}}{\text{h}}}{95\ \frac{\text{kmol}}{\text{h}}}\right) \\ &= 0.774x_N + 0.00526 \end{aligned}$$

There is no vaporization of absorbent into carrier gas, nor any absorption of carrier gas by liquid. So, for the solute at any equilibrium stage N,

$$K_N = \frac{y_N(1 + x_N)}{x_N(1 + y_N)}$$

Rearranging to isolate y_N,

$$\begin{aligned} K_N x_N(1+y_N) &= y_N(1+x_N) \\ K_N x_N + K_N x_N y_N &= y_N + x_N y_N \\ K_N x_N &= y_N + x_N y_N - K_N x_N y_N \\ &= y_N(1 + x_N - K_N x_N) \\ y_N &= \frac{K_N x_N}{1 + x_N - K_N x_N} \\ &= \frac{K_N x_N}{1+(1-K_N)x_N} \end{aligned}$$

Substituting the value of K_N for ethanol gives the equilibrium curve.

$$\begin{aligned} y_N &= \frac{0.573 x_N}{1+(1-0.573)x_N} \\ &= \frac{0.573 x_N}{1+0.427 x_N} \end{aligned}$$

The equilibrium curve and the operating line are drawn through the terminal points (y_1, x_0) and (y_{N+1}, x_N). The number of stages for 90% recovery of ethanol is determined by the mole fraction of ethanol leaving the column in the liquid phase.

$$\begin{aligned} x_N &= \frac{n_{\text{out}}}{L'} \\ &= \frac{4.5 \ \frac{\text{kmol}}{\text{h}}}{73.488 \ \frac{\text{kmol}}{\text{h}}} \\ &= 0.0612 \end{aligned}$$

Use the graphical method to find the number of stages. Start at the top of the column, where y_1 is 0.00526 and x_0 is 0. Locate this point on the operating line and draw a horizontal line to intercept the equilibrium line at the point (0.01240, 0.00526).

From this point, draw a vertical line to intercept the operating line at (0.00923, 0.01240). From this point, draw a horizontal line to intercept the equilibrium curve at (0.02185, 0.01240). Continue alternating between vertical and horizontal lines, reaching the points (0.02185, 0.02215), (0.03933, 0.02215), (0.03933, 0.03567), and (0.06399, 0.03567).

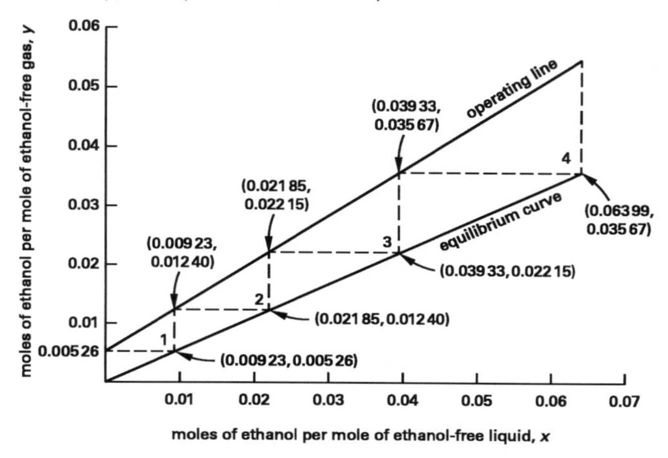

Stage 4 ends with a mole fraction of 0.06399, which is greater than the required mole fraction of 0.0612. So, four (4) stages are required.

The answer is (C).

17. The outside diameter of the pipe is given as

$$D = \frac{100 \text{ mm}}{1000 \ \frac{\text{mm}}{\text{m}}} = 0.100 \text{ m}$$

The absolute temperature of the surface is

$$T_{\text{surface}} = 200^\circ\text{C} + 273^\circ = 473\text{K}$$

The absolute temperature of the surrounding air is

$$T_{\text{air}} = 20^\circ\text{C} + 273^\circ = 293\text{K}$$

The rate of heat loss from the surface is to be calculated per unit length of pipe, so use 1 m for the length of the pipe, L. The surface area of heat transfer is

$$A = \pi D L = \pi(0.100 \text{ m})(1 \text{ m}) = 0.314 \text{ m}^2$$

The heat loss from the pipe to the room air is by convection and by radiation. The convection heat transfer coefficient from the surface of the pipe to the air, h, is 16.5 W/m^2·K. From *NCEES Handbook:* Newton's Law of Cooling, the heat loss per meter of pipe due to convection is

$$\begin{aligned} \dot{Q}_{\text{conv}} &= hA(T_w - T_\infty) \\ &= \left(16.5 \ \frac{\text{W}}{\text{m}^2\cdot\text{K}}\right)(0.314 \text{ m}^2)(473\text{K} - 293\text{K}) \\ &= 932.6 \text{ W} \end{aligned}$$

The emissivity of the pipe surface is given as 0.85. From *NCEES Handbook* table "Physical Constants," the Stefan-Boltzmann constant, σ, is 5.67×10^{-8} W/m²·K⁴. From *NCEES Handbook:* Stefan-Boltzmann Law of Radiation, the Stefan-Boltzmann law of radiation is

$$\dot{Q} = \varepsilon \sigma A T^4$$

Applying the preceding equation to the conditions of the problem statement, the heat loss per meter of pipe due to radiation is

$$\begin{aligned}\dot{Q}_{\text{rad}} &= \varepsilon A \sigma (T_{\text{surface}}^4 - T_{\text{air}}^4) \\ &= (0.85)(0.314\ \text{m}^2)\left(5.67 \times 10^{-8}\ \frac{\text{W}}{\text{m}^2\cdot\text{K}^4}\right) \\ &\quad \times \left((473\text{K})^4 - (293\text{K})^4\right) \\ &= 646.0\ \text{W}\end{aligned}$$

The total heat loss from the pipe to the room air unit length of pipe is

$$\begin{aligned}\dot{Q}_{\text{total}} &= \frac{\dot{Q}_{\text{conv}} + \dot{Q}_{\text{rad}}}{L} \\ &= \frac{932.6\ \text{W} + 646.0\ \text{W}}{1\ \text{m}} \\ &= 1578.6\ \text{W/m} \quad (1580\ \text{W/m})\end{aligned}$$

The answer is (B).

18. The splitting point on the diagram is marked in the illustration shown.

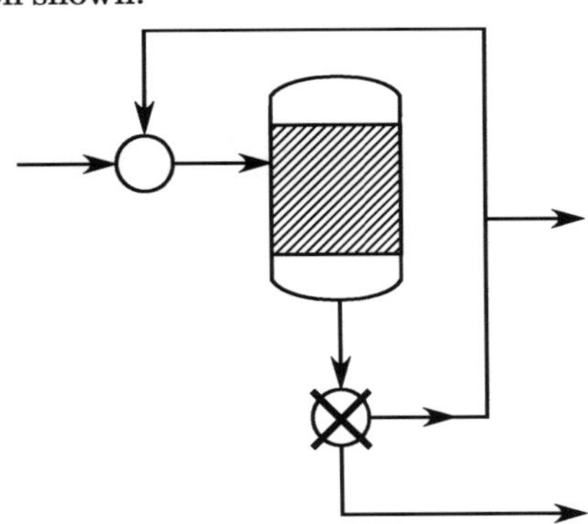

The answer is marked with an X.

19. The inside diameter of the pipe is

$$D = \frac{4.026\ \text{in}}{12\ \frac{\text{in}}{\text{ft}}} = 0.3355\ \text{ft}$$

The cross-sectional area of the pipe is

$$\begin{aligned}A &= \frac{\pi D^2}{4} \\ &= \frac{\pi (0.3355\ \text{ft})^2}{4} \\ &= 0.08840\ \text{ft}^2\end{aligned}$$

The flow rate of the water, $\dot{Q}$, is 50 gal/min. The velocity of the water at the entrance of the scrubber is

$$\begin{aligned}u_{\text{B}} &= \frac{\dot{Q}}{A} \\ &= \frac{50\ \frac{\text{gal}}{\text{min}}}{(0.08840\ \text{ft}^2)\left(7.48\ \frac{\text{gal}}{\text{ft}^3}\right)\left(60\ \frac{\text{sec}}{\text{min}}\right)} \\ &= 1.26\ \text{ft/sec}\end{aligned}$$

The density of the water, ρ, is 62.2 lbm/ft³. The viscosity of the water is

$$\begin{aligned}\mu &= (0.86\ \text{cP})\left(6.72 \times 10^{-4}\ \frac{\frac{\text{lbm}}{\text{ft-sec}}}{\text{cP}}\right) \\ &= 0.0005779\ \text{lbm/ft-sec}\end{aligned}$$

From *NCEES Handbook* table "Dimensionless Numbers," the Reynolds number at the entrance of the scrubber is

$$\begin{aligned}Re &= \frac{\rho u_{\text{B}} D}{\mu} \\ &= \frac{\left(62.2\ \frac{\text{lbm}}{\text{ft}^3}\right)\left(1.26\ \frac{\text{ft}}{\text{sec}}\right)(0.3355\ \text{ft})}{0.0005779\ \frac{\text{lbm}}{\text{ft-sec}}} \\ &= 45{,}499\end{aligned}$$

The roughness of the pipe, ε, is 0.00007 ft. From *NCEES Handbook:* Absolute Roughness and Relative Roughness, the relative roughness of the pipe is

$$\frac{\varepsilon}{D} = \frac{0.00007\ \text{ft}}{0.3355\ \text{ft}} = 0.000209$$

With the Reynolds number and the relative roughness, the Moody diagram is used to find the Darcy friction factor, *f*, which is 0.022. The equivalent length of the pipe, L, is 248 ft. From *NCEES Handbook:* Head Loss in Pipe or Conduit, the Darcy-Weisbach equation is

$$\begin{aligned} h_L &= f\frac{L}{D}\frac{u_B^2}{2g} \\ &= (0.022)\left(\frac{(248\text{ ft})\left(1.26\ \frac{\text{ft}}{\text{sec}}\right)^2}{(0.3355\text{ ft})(2)\left(32.2\ \frac{\text{ft}}{\text{sec}^2}\right)}\right) \\ &= 0.4009\text{ ft} \end{aligned}$$

As in *NCEES Handbook:* The Bernoulli Equation, the Bernoulli equation is

$$\frac{P_A g_c}{\rho g} + \frac{u_A^2}{2g} + z_A + h_s = \frac{P_B g_c}{\rho g} + \frac{u_B^2}{2g} + z_B + h_L + h_v$$

The tank is large, so the velocity of the water on the surface of the tank, u_A, may be considered to be zero. The scrubber and the tank are both at atmospheric pressure, so P_A and P_B are both 14.7 lbf/in^2. The pump head, h_s, is 136 ft. Solving the energy balance equation for the control valve head, h_v, gives

$$\begin{aligned} h_v &= \frac{(P_A - P_B)g_c}{\rho g} + \frac{u_A^2 - u_B^2}{2g} + (z_A - z_B) + h_s - h_L \\ &= \frac{\left(14.7\ \frac{\text{lbf}}{\text{in}^2} - 14.7\ \frac{\text{lbf}}{\text{in}^2}\right)\left(12\ \frac{\text{in}}{\text{ft}}\right)^2\left(32.2\ \frac{\text{ft-lbm}}{\text{lbf-sec}^2}\right)}{\left(62.2\ \frac{\text{lbm}}{\text{ft}^3}\right)\left(32.2\ \frac{\text{ft}}{\text{sec}^2}\right)} \\ &\quad + \frac{\left(0\ \frac{\text{ft}}{\text{sec}}\right)^2 - \left(1.26\ \frac{\text{ft}}{\text{sec}}\right)^2}{(2)\left(32.2\ \frac{\text{ft}}{\text{sec}^2}\right)} + (0\text{ ft} - 65\text{ ft}) \\ &\quad + 136\text{ ft} - 0.4009\text{ ft} \\ &= 70.57\text{ ft} \quad (71\text{ ft}) \end{aligned}$$

The answer is (C).

20. The mixture of recycled air and outside air is cooled and dehumidified by the cooling coil and circulated back to the conditioned space. Outside air makes up 28% of the flow of supply air, so the ratio, r, of mass flow rate of the outside air to the mass flow rate of the supply air mixture is 0.28.

The space is at 75°F and 50% relative humidity. From *NCEES Handbook:* Psychrometric Chart (U.S. Customary Units), the psychrometric chart for normal temperatures, the specific enthalpy of the conditioned space, h_{in}, is 28.1 Btu/lbm dry air. The outside air is at 96°F dry-bulb and 80°F wet-bulb, so from the same chart, the specific enthalpy of the outside air, h_{out}, is 43.4 Btu/lbm dry air. On the chart, the mixture state lies on the straight line drawn between the outside and inside states. An energy balance around the mixer gives the enthalpy of the mixed air as it leaves the mixer.

$$\begin{aligned} h_{mix} &= h_{in} + r(h_{out} - h_{in}) \\ &= 28.1\ \frac{\text{Btu}}{\text{lbm dry air}} + (0.28) \\ &\quad \times \left(43.4\ \frac{\text{Btu}}{\text{lbm dry air}} - 28.1\ \frac{\text{Btu}}{\text{lbm dry air}}\right) \\ &= 32.4\text{ Btu/lbm dry air} \end{aligned}$$

The supply air is at 50°F and 86% relative humidity, so from *NCEES Handbook:* Psychrometric Chart (U.S. Customary Units), its enthalpy, h_{supply}, is 16.1 Btu/lbm dry air. The enthalpy of the supply air rises as it mixes with the air in the conditioned space. To keep the conditioned space at a constant temperature, and because the forced air system is adiabatic, the rate at which the enthalpy of the supply air rises must equal the total cooling load on the conditioned space. This load is given as

$$\dot{Q}_{space} = 65{,}000\text{ Btu/hr}$$

An energy balance around the conditioned space gives

$$\dot{Q}_{space} = \dot{m}_{supply}(h_{in} - h_{supply})$$

Solving for the mass flow rate of the supply air gives

$$\begin{aligned} \dot{m}_{supply} &= \frac{\dot{Q}_{space}}{h_{in} - h_{supply}} \\ &= \frac{65{,}000\ \frac{\text{Btu}}{\text{hr}}}{28.1\ \frac{\text{Btu}}{\text{lbm dry air}} - 16.1\ \frac{\text{Btu}}{\text{lbm dry air}}} \\ &= 5417\text{ lbm dry air/hr} \end{aligned}$$

The mixture of outside and recycled air is cooled and dehumidified by heat transfer to the refrigeration cooling coil. The cooling coil load is

$$\dot{Q}_{\text{coil}} = \dot{m}_{\text{supply}}(h_{\text{mix}} - h_{\text{supply}})$$

$$= \frac{\left(5417 \frac{\text{lbm dry air}}{\text{hr}}\right) \times \left(32.4 \frac{\text{Btu}}{\text{lbm dry air}} - 16.1 \frac{\text{Btu}}{\text{lbm dry air}}\right)}{12{,}000 \frac{\text{Btu}}{\text{ton of refrigeration}}}$$

$$= 7.36 \text{ tons of refrigeration/hr} \quad (7.4 \text{ tons of refrigeration/hr})$$

The answer is (B).

21. The relevant data are shown in the illustration.

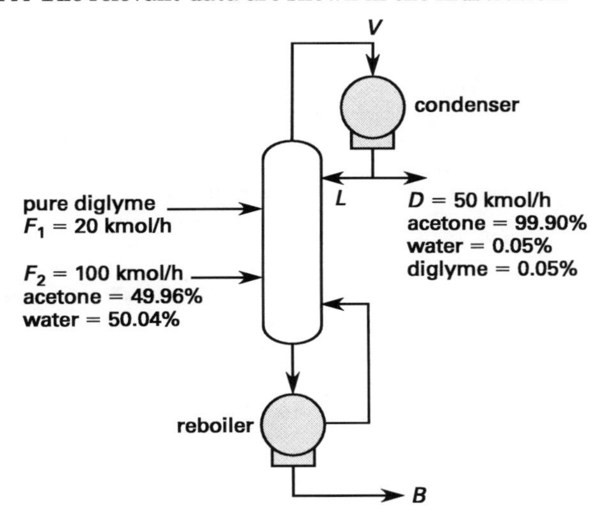

The overall mass balance is (see also *NCEES Handbook:* Material and Energy Balances for Trayed and Packed Units).

$$F = D + B$$

From the overall mass balance around the column, the combined molar flow rate of the two feed streams equals the combined molar flow rate of the distillate and the bottoms stream.

$$F_1 + F_2 = D + B$$

The molar flow rate of the bottoms stream, then, is

$$B = F_1 + F_2 - D = 20 \frac{\text{kmol}}{\text{h}} + 100 \frac{\text{kmol}}{\text{h}} - 50 \frac{\text{kmol}}{\text{h}} = 70 \text{ kmol/h}$$

The feed stream F_2 is 49.96% acetone, so the mole fraction of acetone in the feed, x_F, is 0.4996. The distillate is 99.90% acetone, so the mole fraction of acetone in the distillate, x_D, is 0.9990. The component mass balance is (see also *NCEES Handbook:* Material and Energy Balances for Trayed and Packed Units)

$$x_F F_2 = x_D D + x_B B$$

A mass balance of acetone around the column gives

$$x_F F_2 = x_D D + x_B B$$

Solving for the mole fraction of acetone in the bottoms stream,

$$x_B = \frac{x_F F_2 - x_D D}{B} = \frac{(0.4996)\left(100 \frac{\text{kmol}}{\text{h}}\right) - (0.9990)\left(50 \frac{\text{kmol}}{\text{h}}\right)}{70 \frac{\text{kmol}}{\text{h}}} = 0.000\,143$$

The reflux ratio, R, is given as 3. From *NCEES Handbook:* Column Material Balance, the reflux ratio is

$$R = \frac{L}{D}$$

Rearranging the preceding equation gives the molar flow rate of the stream that is refluxed to the column from the total condenser.

$$L = RD = (3)\left(50 \frac{\text{kmol}}{\text{h}}\right) = 150 \text{ kmol/h}$$

For the rectifying section, the following balance applies.

$$D = V - L$$

Rearranging the preceding equation, the molar flow rate of the vapor stream from the column to the condenser is

$$V = L + D = 150 \frac{\text{kmol}}{\text{h}} + 50 \frac{\text{kmol}}{\text{h}} = 200 \text{ kmol/h}$$

The mole fraction of acetone in the vapor stream reaching plate N is y. The mole fraction of acetone in the liquid stream leaving plate N is x. Because the two feed streams are saturated liquids, they contain no vapor phase. The molar flow rate of the vapor stream reaching stage N is

$$V' = V = 200 \text{ kmol/h}$$

The molar flow rate of the liquid stream leaving stage N is L'. The streams above stage N are

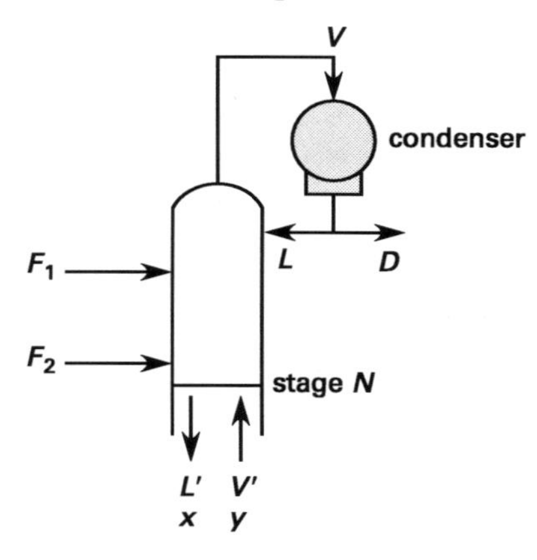

The streams below stage N are

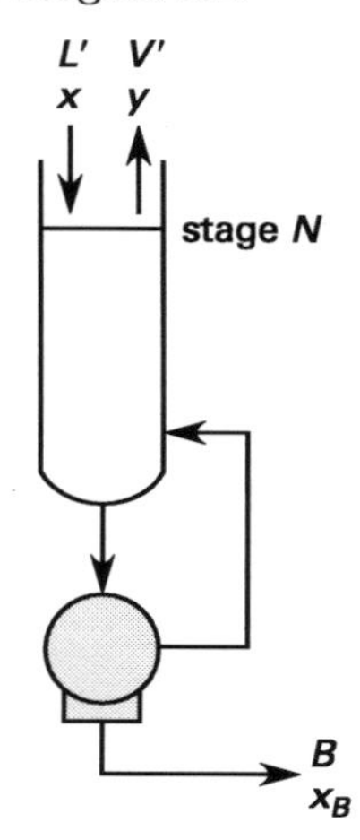

The two feed streams are saturated liquids and, consequently, are 100% liquid. In the stripping section, the molar flow rates of these two saturated liquid streams are added to the molar flow rate of the liquid stream from the reflux.

$$\begin{aligned} L' &= F_1 + F_2 + L \\ &= 20\ \frac{\text{kmol}}{\text{h}} + 100\ \frac{\text{kmol}}{\text{h}} + 150\ \frac{\text{kmol}}{\text{h}} \\ &= 270 \text{ kmol/h} \end{aligned}$$

From *NCEES Handbook:* Material and Energy Balances for Trayed and Packed Units, the component mass balance is

$$x_B B = yV' - xL'$$

A mass balance of acetone around the bottom section gives

$$yV' = xL' + x_B B$$

Solving for the mole fraction of acetone in the vapor stream gives

$$\begin{aligned} y &= \frac{xL' + x_B B}{V'} \\ &= \frac{x\left(270\ \frac{\text{kmol}}{\text{h}}\right) + (0.000143)\left(70\ \frac{\text{kmol}}{\text{h}}\right)}{200\ \frac{\text{kmol}}{\text{h}}} \\ &= 1.35x + 0.000050 \end{aligned}$$

The answer is (D).

22. The relevant data are shown in the illustration.

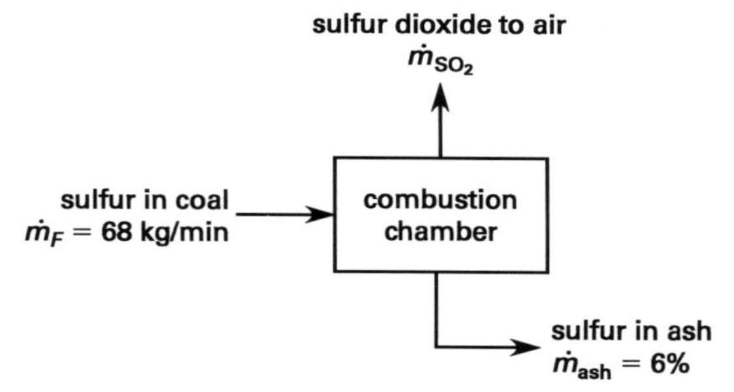

The coal contains 3.5% sulfur (by weight), so the sulfur flow rate in the coal is

$$\begin{aligned} \dot{m}_S &= 0.035\dot{m}_F \\ &= (0.035)\left(68\ \frac{\text{kg}}{\text{min}}\right)\left(60\ \frac{\text{min}}{\text{h}}\right)\left(24\ \frac{\text{h}}{\text{d}}\right)\left(365\ \frac{\text{d}}{\text{yr}}\right) \\ &= 1\,250\,928 \text{ kg/yr} \end{aligned}$$

The sulfur in the ash is 6% (by weight) of the input sulfur, so the sulfur in the ash is

$$\begin{aligned} \dot{m}_{\text{ash}} &= 0.06\dot{m}_S \\ &= (0.06)\left(1\,250\,928\ \frac{\text{kg}}{\text{yr}}\right) \\ &= 75\,056 \text{ kg/yr} \end{aligned}$$

From *NCEES Handbook:* Material Balances, the balance equation is

$$\begin{aligned} \textit{Accumulation} &= \textit{Input} - \textit{Output} \\ &\quad + \textit{Generation} - \textit{Consumption} \\ &= \textit{Input} - \textit{Output} + 0 - 0 \end{aligned}$$

The mass flow rate of sulfur available for conversion to SO_2 is

$$\begin{aligned}\dot{m}_{SO_2,\text{avail}} &= \dot{m}_S - \dot{m}_{\text{ash}} \\ &= 1\,250\,928\ \frac{\text{kg}}{\text{yr}} - 75\,056\ \frac{\text{kg}}{\text{yr}} \\ &= 1\,175\,872\ \text{kg/yr}\end{aligned}$$

The molecular weight of sulfur, MW_S, is given as 32 kg/kmol. The molecular weight of sulfur dioxide, MW_{SO_2}, is given as 64 kg/kmol. The amount of sulfur dioxide formed is determined from the proportional weights of the oxidation reaction.

$$S + O_2 \rightarrow SO_2$$

The mass flow rate of sulfur dioxide formed is

$$\begin{aligned}\dot{m}_{SO_2} &= \left(\frac{MW_{SO_2}}{MW_S}\right)\dot{m}_{SO_2,\text{avail}} \\ &= \left(\frac{64\ \frac{\text{kg}}{\text{kmol}}}{32\ \frac{\text{kg}}{\text{kmol}}}\right)\left(1\,175\,872\ \frac{\text{kg}}{\text{yr}}\right) \\ &= 2\,351\,744\ \text{kg/yr} \quad (2.4 \times 10^6\ \text{kg/yr})\end{aligned}$$

The answer is (C).

23. The predicted landfill gas production rate, LFG, is 3.60×10^6 m^3/yr. Because the landfill gas is 60% methane, the annual predicted volumetric flow rate of methane is

$$\begin{aligned}\dot{V} &= 0.60(\text{LFG}) \\ &= (0.60)\left(3.60 \times 10^6\ \frac{\text{m}^3}{\text{yr}}\right) \\ &= 2.16 \times 10^6\ \text{m}^3/\text{yr}\end{aligned}$$

The heating value, HV, of pure methane is 33 810 kJ/m^3. The heating capacity of the landfill gas is

$$\begin{aligned}\dot{Q} &= (\text{HV})\,\dot{V} \\ &= \left(33\,810\ \frac{\text{kJ}}{\text{m}^3}\right)\left(2.16 \times 10^6\ \frac{\text{m}^3}{\text{yr}}\right) \\ &= 7.303 \times 10^{10}\ \text{kJ/yr}\end{aligned}$$

The engine's rate of energy conversion is

$$C = 14\,235\ \text{kJ/kW·h}$$

The landfill generating capacity is

$$\begin{aligned}\text{LC} &= \frac{\dot{Q}}{C} \\ &= \frac{7.303 \times 10^{10}\ \frac{\text{kJ}}{\text{yr}}}{14\,235\ \frac{\text{kJ}}{\text{kW·h}}} \\ &= 5.13 \times 10^6\ \text{kW·h/yr}\end{aligned}$$

The selling price of the electricity is given as

$$\text{EC} = \$0.08/\text{kW·h}$$

The annual value of the landfill generating capacity is

$$\begin{aligned}A &= (\text{LC})(\text{EC})(1\ \text{yr}) \\ &= \left(5.13 \times 10^6\ \frac{\text{kW·h}}{\text{yr}}\right)\left(\frac{\$0.08}{\text{kW·h}}\right)(1\ \text{yr}) \\ &= \$410{,}400\end{aligned}$$

The number of periods (years) that the landfill is being designed for is

$$n = 20$$

The expected interest rate per period is

$$i = 0.07$$

From *NCEES Handbook:* Cost Estimation and Project Evaluation, the present value of the investment is

$$\begin{aligned}P &= A\left(\frac{(1+i)^n - 1}{i(1+i)^n}\right) \\ &= (\$410{,}400)\left(\frac{(1+0.07)^{20} - 1}{(0.07)(1+0.07)^{20}}\right) \\ &= \$4{,}347{,}783 \quad (\$4.0\ \text{million})\end{aligned}$$

The answer is (B).

24. The liquid mass flow rates into and out of both tanks are equal.

$$\dot{m} = 210\ \text{kg/min}$$

The mass of the liquid in each tank therefore remains constant.

$$m = 2300\ \text{kg}$$

The liquid enters the first tank at an initial temperature, T_0, of 25°C. The temperature of the liquid in each tank increases until it reaches steady state. Let T_1 be the liquid's steady-state temperature when leaving the first tank and entering the second tank, and let T_2 be its steady-state temperature when leaving the second tank. From *NCEES Handbook:* Combination of Heat-Transfer Mechanisms, the rate of heat transfer is

$$\dot{Q} = U_{ov}A\Delta T$$

The rate of heat transfer through the coils to the liquid in the first tank is

$$\dot{Q} = U_{ov}A(T_{steam} - T_1)$$

The overall heat transfer coefficient, U_{ov}, is given as 15 kJ/m^2·min·°C. The area of heat transfer, A, is given as 1 m^2. The temperature of the steam in the coils in each of the tanks, T_{steam}, is given as 320°C.

From *NCEES Handbook:* Heat Capacity/Specific Heat (c_p), the heat transferred is

$$\dot{Q} = mc_p\frac{dT}{dt}$$

Perform an energy balance for the first tank. The total heat added to the liquid in the first tank equals the heat that enters the tank with the incoming liquid plus the heat that is transferred to the liquid from the steam, minus the heat that leaves the tank with the exiting liquid.

$$mc_p\frac{dT_1}{dt} = \dot{m}c_pT_0 + \dot{Q} - \dot{m}c_pT_1$$

From *NCEES Handbook:* Combination of Heat-Transfer Mechanisms,

$$\dot{Q} = U_{ov}A\Delta T$$

Or,

$$mc_p\frac{dT_1}{dt} = \dot{m}c_pT_0 + U_{ov}A(T_{steam} - T_1) - \dot{m}c_pT_1$$

The heat capacity of the liquid, c_p, is given as 2.03 kJ/kg·°C. At steady state, the temperature change over time is zero.

$$\dot{m}c_pT_0 + U_{ov}A(T_{steam} - T_1) - \dot{m}c_pT_1 = 0\ \frac{°\text{C}}{\text{min}}$$

Solving for T_1,

$$\begin{aligned} T_1 &= \frac{\dot{m}c_pT_0 + U_{ov}AT_{steam}}{\dot{m}c_p + U_{ov}A} \\ &= \frac{\left(\begin{array}{l}\left(210\ \frac{\text{kg}}{\text{min}}\right)\left(2.03\ \frac{\text{kJ}}{\text{kg·°C}}\right)(25°\text{C}) \\ \quad + \left(15\ \frac{\text{kJ}}{\text{m}^2\text{·min·°C}}\right)(1\ \text{m}^2)(320°\text{C})\end{array}\right)}{\left(\begin{array}{l}\left(210\ \frac{\text{kg}}{\text{min}}\right)\left(2.03\ \frac{\text{kJ}}{\text{kg·°C}}\right) \\ \quad + \left(15\ \frac{\text{kJ}}{\text{m}^2\text{·min·°C}}\right)(1\ \text{m}^2)\end{array}\right)} \\ &= 35.03°\text{C} \end{aligned}$$

Similarly, the rate of heat transfer through the coils to the liquid in the second tank is

$$\dot{Q} = U_{ov}A(T_{steam} - T_2)$$

A similar energy balance for the second tank gives

$$mc_p\frac{dT_2}{dt} = \dot{m}c_pT_1 + U_{ov}A(T_{steam} - T_2) - \dot{m}c_pT_2$$

At steady state, the temperature change over time is zero.

$$\dot{m}c_pT_1 + U_{ov}A(T_{steam} - T_2) - \dot{m}c_pT_2 = 0\ \frac{°\text{C}}{\text{min}}$$

Solving for T_2,

$$\begin{aligned} T_2 &= \frac{\dot{m}c_pT_1 + U_{ov}AT_{steam}}{\dot{m}c_p + U_{ov}A} \\ &= \frac{\left(\begin{array}{l}\left(210\ \frac{\text{kg}}{\text{min}}\right)\left(2.03\ \frac{\text{kJ}}{\text{kg·°C}}\right)(35.03°\text{C}) \\ \quad + \left(15\ \frac{\text{kJ}}{\text{m}^2\text{·min·°C}}\right)(1\ \text{m}^2)(320°\text{C})\end{array}\right)}{\left(\begin{array}{l}\left(210\ \frac{\text{kg}}{\text{min}}\right)\left(2.03\ \frac{\text{kJ}}{\text{kg·°C}}\right) \\ \quad + \left(15\ \frac{\text{kJ}}{\text{m}^2\text{·min·°C}}\right)(1\ \text{m}^2)\end{array}\right)} \\ &= 44.72°\text{C} \quad (45°\text{C}) \end{aligned}$$

The answer is (C).

25. The weight fraction of water in the product, x, is 0.074. The mass flow rate of the water in the product leaving the dryer is

$$\dot{m}_{\text{water,out}} = x\dot{m}_{\text{product,out}} = (0.074)\left(345\ \frac{\text{kg}}{\text{h}}\right) = 25.53\ \text{kg/h}$$

From *NCEES Handbook:* Material Balances With No Reaction, the balance equation is

Accumulation = Input − Output + Generation − Consumption

Because the product consists only of water and dry polymer, the mass balance in the product gives the mass flow rate of dry polymer in the product.

$$\begin{aligned}\dot{m}_{\text{product,out}} &= \dot{m}_{\text{dry polymer,out}} + \dot{m}_{\text{water,out}}\\ \dot{m}_{\text{dry polymer,out}} &= \dot{m}_{\text{product,out}} - \dot{m}_{\text{water,out}}\\ &= 345\ \frac{\text{kg}}{\text{h}} - 25.53\ \frac{\text{kg}}{\text{h}}\\ &= 319.47\ \text{kg/h}\end{aligned}$$

The dryer affects only the water in the feed, so the dry polymer has the same mass at the entrance and exit of the dryer.

$$\dot{m}_{\text{dry polymer,in}} = \dot{m}_{\text{dry polymer,out}} = 319.47\ \text{kg/h}$$

The ratio of water to dry polymer at the entrance of the dryer is given as

$$r = 1.43\ \text{g water/g dry polymer}$$

The mass flow rate of water at the entrance of the dryer is

$$\begin{aligned}\dot{m}_{\text{water,in}} &= r\dot{m}_{\text{solid,in}}\\ &= \left(1.43\ \frac{\text{g water}}{\text{g dry polymer}}\right)\\ &\quad\times\left(\left(319.47\ \frac{\text{kg dry polymer}}{\text{h}}\right)\left(1000\ \frac{\text{g}}{\text{kg}}\right)\right)\\ &= 456\,842.1\ \text{g water/h}\end{aligned}$$

The mass of water in the product entering the dryer equals the mass of water in the product leaving the dryer plus the water in the air leaving the dryer. The mass balance of water around the dryer gives

$$\begin{aligned}\dot{m}_{\text{water,in}} &= \dot{m}_{\text{water,out}} + \dot{m}_{\text{water,air}}\\ \dot{m}_{\text{water,air}} &= \dot{m}_{\text{water,in}} - \dot{m}_{\text{water,out}}\\ &= 456\,842.1\ \frac{\text{g}}{\text{h}} - \left(25.53\ \frac{\text{kg}}{\text{h}}\right)\left(1000\ \frac{\text{g}}{\text{kg}}\right)\\ &= 431\,312.1\ \text{g/h}\end{aligned}$$

The product temperature at the entrance to the dryer is 22°C, and the temperature at the exit is 43°C. The heat capacity of the water, $c_{p,\text{water}}$, is given as 4.184 J/g·°C. From *NCEES Handbook:* Heat Capacity/Specific Heat (c_p), the enthalpy change of the water removed by the air is

$$\begin{aligned}\dot{Q} &= \dot{m}c_p\Delta T\\ \dot{Q}_{\text{water}} &= \dot{m}_{\text{water,air}}c_{p,\text{water}}(T_{\text{out}} - T_{\text{in}})\\ &= \left(431\,312.1\ \frac{\text{g}}{\text{h}}\right)\left(4.184\ \frac{\text{J}}{\text{g}\cdot{}^\circ\text{C}}\right)(43^\circ\text{C} - 22^\circ\text{C})\\ &= 37\,896\,806.35\ \text{J/h}\end{aligned}$$

The heat capacity of the dry polymer, $c_{p,\text{dry polymer}}$, is given as 0.36 J/g·°C. From *NCEES Handbook:* Heat Capacity/Specific Heat (c_p), the enthalpy change of the dry polymer is

$$\begin{aligned}\dot{Q} &= \dot{m}c_p\Delta T\\ \dot{Q}_{\text{dry polymer}} &= \dot{m}_{\text{dry polymer,in}}c_{p,\text{dry polymer}}(T_{\text{out}} - T_{\text{in}})\\ &= \left(\left(319.47\ \frac{\text{kg}}{\text{h}}\right)\left(1000\ \frac{\text{g}}{\text{kg}}\right)\right)\\ &\quad\times\left(0.36\ \frac{\text{J}}{\text{g}\cdot{}^\circ\text{C}}\right)(43^\circ\text{C} - 22^\circ\text{C})\\ &= 2\,415\,193.2\ \text{J/h}\end{aligned}$$

The total enthalpy change is

$$\begin{aligned}\dot{Q} &= \dot{Q}_{\text{water}} + \dot{Q}_{\text{dry polymer}}\\ &= 37\,896\,806.35\ \frac{\text{J}}{\text{h}} + 2\,415\,193.2\ \frac{\text{J}}{\text{h}}\\ &= 40\,311\,999.55\ \text{J/h}\quad (4.0\times 10^7\ \text{J/h})\end{aligned}$$

The answer is (D).

26. The heat generated within the wall is transferred to the water because the outer surface of the wall is insulated. The rate of energy generated must equal the rate at which the heat is transferred from the tube to the water flowing in the tube. The amount of heat generated, q, is $1.5\times 10^6\ \text{W/m}^3$.

The energy transfer through the wall is

$$E = q\left(\frac{\pi}{4}\right)(D_o^2 - D_i^2)L$$

Because the heat is uniform, the heat transferred through the surface of the tube is

$$\dot{Q}_s = \frac{E}{\pi D_i L} = \frac{q\left(\frac{\pi}{4}\right)(D_o^2 - D_i^2)L}{\pi D_i L} = \frac{q(D_o^2 - D_i^2)}{4D_i}$$

$$= \frac{\left(1.5\times 10^6\ \frac{\text{W}}{\text{m}^3}\right)\left(\left(\frac{60\text{mm}}{1000\ \frac{\text{mm}}{\text{m}}}\right)^2 - \left(\frac{40\text{mm}}{1000\ \frac{\text{mm}}{\text{m}}}\right)^2\right)}{(4)\left(\frac{40\text{mm}}{1000\ \frac{\text{mm}}{\text{m}}}\right)}$$

$$= 18\,750\ \text{W/m}^2$$

From *NCEES Handbook:* Convection, Newton's law of cooling is

$$\dot{Q} = hA(T_w - T_\infty)$$

From Newton's law of cooling, the convection coefficient, per unit area, at the tube exit is

$$h = \frac{\dot{Q}_s}{T_w - T_\infty} = \frac{18\,750\ \frac{\text{W}}{\text{m}^2}}{76°\text{C} - 68°\text{C}}$$

$$= \frac{18\,750\ \frac{\text{W}}{\text{m}^2}}{8°\text{C}}$$

A temperature difference of 1°C is equal to a temperature difference of 1K, so

$$h = \frac{18\,750\ \frac{\text{W}}{\text{m}^2}}{8\text{K}}$$

$$= 2343.75\ \text{W/m}^2\text{·K} \quad (2300\ \text{W/m}^2\text{·K})$$

The answer is (D).

27. The cyclone separator is marked in the illustration shown.

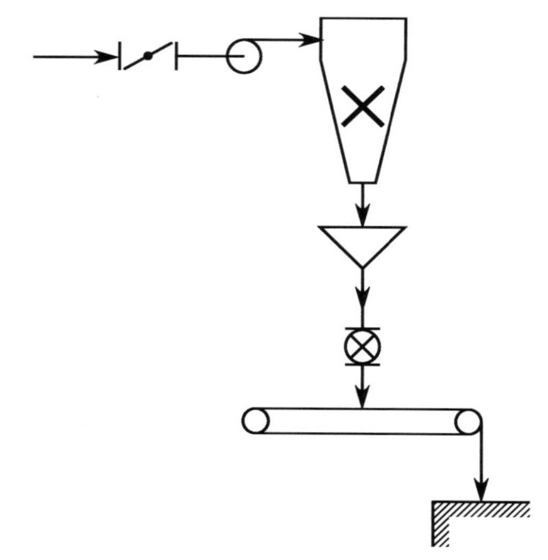

The answer is marked with an X.

28. The inside diameter of the pipe is

$$D = \frac{1.38\ \text{in}}{12\ \frac{\text{in}}{\text{ft}}} = 0.115\ \text{ft}$$

The cross-sectional area of the pipe is

$$A = \frac{\pi D^2}{4}$$

$$= \frac{\pi(0.115\ \text{ft})^2}{4}$$

$$= 0.010387\ \text{ft}^2$$

The velocity of the fluid in the pipe (point A) is

$$u_\text{A} = \frac{Q}{A}$$

$$= \frac{\frac{32\ \frac{\text{gal}}{\text{min}}}{\left(7.48\ \frac{\text{gal}}{\text{ft}^3}\right)\left(60\ \frac{\text{sec}}{\text{min}}\right)}}{0.010387\ \text{ft}^2}$$

$$= 6.864\ \text{ft/sec}$$

The tank and the pipe are open to the atmosphere, so the pressure head, $p_\text{B} - p_\text{A}$, is zero. The liquid flows by gravity, so the velocity on the surface of the fluid (point B), u_B, is zero. The system does not involve a pump, so there is no added head.

The density of the fluid, ρ, is 62.22 lbm/ft^3. The viscosity of the fluid is given as

$$\mu = (0.743 \text{ cP})\left(6.72\times 10^{-4}\ \frac{\frac{\text{lbm}}{\text{ft-sec}}}{\text{cP}}\right) = 0.0004993 \text{ lbm/ft-sec}$$

From *NCEES Handbook* table "Dimensionless Numbers," the Reynolds number is

$$\begin{aligned} Re &= \frac{\rho u_A D}{\mu} \\ &= \frac{\left(62.22\ \frac{\text{lbm}}{\text{ft}^3}\right)\left(6.864\ \frac{\text{ft}}{\text{sec}}\right)(0.115 \text{ ft})}{0.0004993\ \frac{\text{lbm}}{\text{ft-sec}}} \\ &= 98{,}365.7 \end{aligned}$$

The specific roughness of the pipe, ε, is 0.00015 ft. From *NCEES Handbook:* Absolute Roughness and Relative Roughness, the relative roughness of the pipe is

$$\frac{\varepsilon}{D} = \frac{0.00015 \text{ ft}}{0.115 \text{ ft}} = 0.001304$$

Using the Reynolds number and the relative roughness of the pipe, the friction factor, f, is found from the Moody chart to be 0.0232. The equivalent length of the 90° elbow is found from the diameter and the equivalent length to diameter ratio.

$$\begin{aligned} L_{\text{eq,elbow}} &= D\left(\frac{L_{\text{eq}}}{D}\right)_{\text{elbow}} \\ &= (0.115 \text{ ft})(30) \\ &= 3.45 \text{ ft} \end{aligned}$$

The equivalent length of the gate valve is

$$\begin{aligned} L_{\text{eq,gate}} &= D\left(\frac{L_{\text{eq}}}{D}\right)_{\text{gate}} \\ &= (0.115 \text{ ft})(160) \\ &= 18.4 \text{ ft} \end{aligned}$$

The equivalent length of the globe valve is

$$\begin{aligned} L_{\text{eq,globe}} &= D\left(\frac{L_{\text{eq}}}{D}\right)_{\text{globe}} \\ &= (0.115 \text{ ft})(340) \\ &= 39.1 \text{ ft} \end{aligned}$$

The equivalent length of the heat exchanger, $L_{\text{eq,exch}}$, is given as 40 ft. The length of the straight pipe (not including the distances covered by the fittings), L_{pipe}, is given as 51 ft. The equivalent length of the horizontal pipe (not including the vertical portion) is

$$\begin{aligned} L_{\text{eq,horiz}} &= L_{\text{eq,elbow}} + L_{\text{eq,gate}} + L_{\text{eq,globe}} + L_{\text{eq,exch}} + L_{\text{pipe}} \\ &= 3.45 \text{ ft} + 18.4 \text{ ft} + 39.1 \text{ ft} + 40 \text{ ft} + 51 \text{ ft} \\ &= 151.95 \text{ ft} \end{aligned}$$

The total equivalent length of the pipe is

$$L_{\text{eq,total}} = L_{\text{eq,horiz}} + L_{\text{vert}}$$

From *NCEES Handbook:* Head Loss in Pipe or Conduit, the Darcy-Weisbach equation is

$$h_L = f\frac{L_{\text{eq,total}}}{D}\frac{u_A^2}{2g}$$

The static head is

$$z_B - z_A = h + L_{\text{vert}}$$

As in *NCEES Handbook:* The Bernoulli Equation, a general mechanical energy balance for flow from point A to point B with no pump work and friction is

$$\frac{P_B g_c}{\rho g} + \frac{u_B^2}{2g} + z_B = \frac{P_A g_c}{\rho g} + \frac{u_A^2}{2g} + z_A + h_L$$

$$\frac{(P_B - P_A)g_c}{\rho g} + \frac{u_B^2}{2g} + (z_B - z_A) = \frac{u_A^2}{2g} + h_L$$

$$\frac{(0 \text{ ft})g_c}{\rho g} + \frac{\left(0\ \frac{\text{ft}}{\text{sec}}\right)^2}{2g} + (h + L_{\text{vert}}) = \frac{u_A^2}{2g} + f\frac{L_{\text{eq,total}}}{D}\frac{u_A^2}{2g}$$

$$h + L_{\text{vert}} = \frac{u_A^2}{2g} + f\frac{L_{\text{eq,horiz}} + L_{\text{vert}}}{D}\frac{u_A^2}{2g}$$

Rearranging and solving for L_{vert},

$$\begin{aligned} L_{\text{vert}} &= \frac{u_A^2}{2g} + f\frac{L_{\text{eq,horiz}} + L_{\text{vert}}}{D}\frac{u_A^2}{2g} - h \\ &= \frac{u_A^2}{2g} + f\frac{L_{\text{eq,horiz}}}{D}\frac{u_A^2}{2g} + f\frac{L_{\text{vert}}}{D}\frac{u_A^2}{2g} - h \\ &= \frac{\left(6.864\ \frac{\text{ft}}{\text{sec}}\right)^2}{(2)\left(32.2\ \frac{\text{ft}}{\text{sec}^2}\right)} \\ &\quad + (0.0232)\left(\frac{151.95\ \text{ft}}{0.115\ \text{ft}}\right)\left(\frac{\left(6.864\ \frac{\text{ft}}{\text{sec}}\right)^2}{(2)\left(32.2\ \frac{\text{ft}}{\text{sec}^2}\right)}\right) \\ &\quad + (0.0232)\left(\frac{L_{\text{vert}}}{0.115\ \text{ft}}\right)\left(\frac{\left(6.864\ \frac{\text{ft}}{\text{sec}}\right)^2}{(2)\left(32.2\ \frac{\text{ft}}{\text{sec}^2}\right)}\right) - 7\ \text{ft} \\ &= 16.16\ \text{ft} + 0.1476 L_{\text{vert}} \\ &= \frac{16.16\ \text{ft}}{1 - 0.1476} \\ &= 18.96\ \text{ft} \end{aligned}$$

The height of the level of the surface of the liquid with respect to the pipe system is

$$\begin{aligned} z_B - z_A &= h + L_{\text{vert}} \\ z_B - 0\ \text{ft} &= 7\ \text{ft} + 18.96\ \text{ft} \\ z_B &= 25.96\ \text{ft} \quad (26\ \text{ft}) \end{aligned}$$

The answer is (D).

29. The diameter of the pipe is given as

$$D = \frac{2.067\ \text{in}}{12\ \frac{\text{in}}{\text{ft}}} = 0.172\ \text{ft}$$

The cross-sectional area of the pipe is

$$\begin{aligned} A &= \frac{\pi D^2}{4} \\ &= \frac{\pi (0.172\ \text{ft})^2}{4} \\ &= 0.0232\ \text{ft}^2 \end{aligned}$$

The water flow rate is given as

$$\dot{Q} = \frac{85\ \frac{\text{gal}}{\text{min}}}{7.48\ \frac{\text{gal}}{\text{ft}^3}} = 11.36\ \text{ft}^3/\text{min}$$

The velocity of the water is

$$\begin{aligned} u &= \frac{\dot{Q}}{A} \\ &= \frac{\frac{11.36\ \frac{\text{ft}^3}{\text{min}}}{60\ \frac{\text{sec}}{\text{min}}}}{0.0232\ \text{ft}^2} \\ &= 8.16\ \text{ft/sec} \end{aligned}$$

The density of the water is given as

$$\rho = 62.2\ \text{lbm/ft}^3$$

The viscosity of the water is

$$\begin{aligned} \mu &= (0.86\ \text{cP})\left(6.72 \times 10^{-4}\ \frac{\frac{\text{lbm}}{\text{ft-sec}}}{\text{cP}}\right) \\ &= 0.000578\ \text{lbm/ft-sec} \end{aligned}$$

From *NCEES Handbook* table "Dimensionless Numbers," the Reynolds number is

$$\begin{aligned} Re &= \frac{\rho u D}{\mu} \\ &= \frac{\left(62.2\ \frac{\text{lbm}}{\text{ft}^3}\right)\left(8.16\ \frac{\text{ft}}{\text{sec}}\right)(0.172\ \text{ft})}{0.000578\ \frac{\text{lbm}}{\text{ft-sec}}} \\ &= 151{,}000 \end{aligned}$$

The specific roughness of the pipe is given as

$$\varepsilon = 0.00018\ \text{ft}$$

From *NCEES Handbook:* Absolute Roughness and Relative Roughness, the relative roughness of the pipe is

$$\frac{\varepsilon}{D} = \frac{0.00018\ \text{ft}}{0.172\ \text{ft}} = 0.00105$$

With the Reynolds number and the relative roughness determined, the Moody diagram gives the friction factor.

$$f = 0.0216$$

From the pipe system diagram and the given data, the total length of pipe is

$$L = 15 \text{ ft} + 40 \text{ ft} + 300 \text{ ft} + 100 \text{ ft} = 455 \text{ ft}$$

For this length of pipe, from *NCEES Handbook:* Head Loss in Pipe or Conduit, the resistance coefficient is

$$\begin{aligned} K_{\text{pipe}} &= \frac{fL}{D} \\ &= \frac{(0.0216)(455 \text{ ft})}{0.172 \text{ ft}} \\ &= 57.14 \end{aligned}$$

The pipe has four standard 90° threaded elbows each with a resistance coefficient of 0.57, so the resistance coefficient of the elbows is

$$\begin{aligned} K_{\text{elbows}} &= (4)(0.57) \\ &= 2.28 \end{aligned}$$

The resistance coefficient of the check valve, K_{check}, is 28. The resistance coefficient of the gate valve, K_{gate}, is 0.15. The resistance coefficient of the sudden exit, K_{exit}, is 1.0. From *NCEES Handbook:* Head Loss in Pipe or Conduit, the total resistance coefficient for the pipe system is

$$\begin{aligned} \sum K &= K_{\text{pipe}} + K_{\text{elbows}} + K_{\text{check}} + K_{\text{gate}} + K_{\text{exit}} \\ &= 57.14 + 2.28 + 28 + 0.15 + 1.0 \\ &= 88.57 \end{aligned}$$

From *NCEES Handbook:* Head Loss in Pipe or Conduit, the head loss through the pipe is

$$\begin{aligned} h &= \frac{(\sum K)u^2}{2g} \\ &= \frac{(88.57)\left(8.16 \ \frac{\text{ft}}{\text{sec}}\right)^2}{(2)\left(32.2 \ \frac{\text{ft}}{\text{sec}^2}\right)} \\ &= 91.6 \text{ ft} \end{aligned}$$

From the illustration, the level of the exit with respect to the ground level is

$$\begin{aligned} h_L &= 300 \text{ ft} + 100 \text{ ft} \\ &= 400 \text{ ft} \end{aligned}$$

The total discharge head of the pump is

$$\begin{aligned} h_{\text{pump}} &= (h_L + h) \times \frac{g}{g_c} \\ &= (400 \text{ ft} + 91.6 \text{ ft})\left(\frac{32.2 \ \frac{\text{ft}}{\text{sec}^2}}{32.2 \ \frac{\text{ft-lbm}}{\text{lbf-sec}^2}}\right) \\ &= 491.6 \text{ ft-lbf/lbm} \end{aligned}$$

The pump efficiency is

$$\eta_{\text{pump}} = 0.75$$

From *NCEES Handbook:* Pump Power, the brake horsepower required is

$$\begin{aligned} BP &= \frac{WP}{\eta_{\text{pump}}} = \frac{\dot{Q}h_{\text{pump}}\rho}{\eta_{\text{pump}}} \\ &= \frac{\left(11.36 \ \frac{\text{ft}^3}{\text{min}}\right)\left(491.6 \ \frac{\text{ft-lbf}}{\text{lbm}}\right)\left(62.2 \ \frac{\text{lbm}}{\text{ft}^3}\right)}{(0.75)\left(33{,}000 \ \frac{\text{ft-lbf}}{\text{hp-min}}\right)} \\ &= 14.03 \text{ hp} \quad (14 \text{ hp}) \end{aligned}$$

The answer is (D).

30. Any ideal gas at standard conditions of 492°R and 14.7 lbf/in^2 has a molar specific volume of 359 ft^3/lb mole. From *NCEES Handbook:* Ideal Gas Law, for an ideal gas,

$$\frac{P_1}{P_2}\frac{\hat{v}_1}{\hat{v}_2} = \frac{T_1}{T_2}$$

So that for an ideal gas, the relationship $P\hat{v}/T$ remains constant under all conditions.

The actual operating pressure is 15 lbf/in^2, and the actual operating temperature is

$$T_a = 70°\text{F} + 460° = 530°\text{R}$$

In comparing ideal conditions with the actual conditions, the ideal gas law states

$$\frac{P_i \hat{\nu}_i}{T_i} = \frac{P_a \hat{\nu}_a}{T_a}$$

$$\hat{\nu}_a = \frac{P_i \hat{\nu}_i T_a}{T_i P_a} = \frac{\left(14.7\ \frac{\text{lbf}}{\text{in}^2}\right)\left(359\ \frac{\text{ft}^3}{\text{lb mole}}\right)(530^\circ\text{R})}{(492^\circ\text{R})\left(15\ \frac{\text{lbf}}{\text{in}^2}\right)} = 379\ \text{ft}^3/\text{lb mole}$$

The volumetric flow rate of the air at standard conditions is given as

$$\dot{V}_{\text{std}} = 3700\ \text{ft}^3/\text{min}$$

The molar airflow rate at the operating conditions is

$$G = \frac{\dot{V}_{\text{std}}}{\hat{\nu}_a} = \frac{\left(3700\ \frac{\text{ft}^3}{\text{min}}\right)\left(60\ \frac{\text{min}}{\text{hr}}\right)}{379\ \frac{\text{ft}^3}{\text{lb mole}}} = 585.75\ \text{lb mole/hr}$$

The wastewater stream volumetric flow rate is given as

$$L' = \frac{\left(550\ \frac{\text{gal}}{\text{min}}\right)\left(60\ \frac{\text{min}}{\text{hr}}\right)}{7.48\ \frac{\text{gal}}{\text{ft}^3}} = 4412\ \text{ft}^3/\text{hr}$$

The molecular weight of the wastewater is given as

$$MW = 18.02\ \text{lbm/lb mole}$$

The density of the wastewater is given as

$$\rho = 62.4\ \text{lbm/ft}^3$$

The wastewater stream molar flow rate is

$$L = \frac{L'\rho}{MW} = \frac{\left(4412\ \frac{\text{ft}^3}{\text{hr}}\right)\left(62.4\ \frac{\text{lbm}}{\text{ft}^3}\right)}{18.02\ \frac{\text{lbm}}{\text{lb mole}}} = 15{,}278\ \text{lb mole/hr}$$

From *NCEES Handbook* table "Absorption and Stripping," the stripping factor is

$$S = \frac{KG}{L}$$

The stripping factor for C_6H_6 is

$$S_{C_6H_6} = \frac{K_{C_6H_6} G}{L} = \frac{(256)\left(585.75\ \frac{\text{lb mole}}{\text{hr}}\right)}{15{,}278\ \frac{\text{lb mole}}{\text{hr}}} = 9.81$$

The stripping factor for C_7H_8 is

$$S_{C_7H_8} = \frac{K_{C_7H_8} G}{L} = \frac{(248)\left(585.75\ \frac{\text{lb mole}}{\text{hr}}\right)}{15{,}278\ \frac{\text{lb mole}}{\text{hr}}} = 9.51$$

The stripping factor for C_8H_{10} is

$$S_{C_8H_{10}} = \frac{K_{C_8H_{10}} G}{L} = \frac{(285)\left(585.75\ \frac{\text{lb mole}}{\text{hr}}\right)}{15{,}278\ \frac{\text{lb mole}}{\text{hr}}} = 10.93$$

The number of theoretical stages, N, is 3. Because the concentration of the VOCs in the wastewater is small, the Kremser equation applies. For C_6H_6, the fraction stripped is

$$\begin{aligned} f_{C_6H_6} &= \frac{S_{C_6H_6}^{N+1} - S_{C_6H_6}}{S_{C_6H_6}^{N+1} - 1} \\ &= \frac{9.81^{(3+1)} - 9.81}{9.81^{(3+1)} - 1} \\ &= 0.99905 \end{aligned}$$

For C_7H_8, the fraction stripped is

$$\begin{aligned} f_{C_7H_8} &= \frac{S_{C_7H_8}^{N+1} - S_{C_7H_8}}{S_{C_7H_8}^{N+1} - 1} \\ &= \frac{9.51^{(3+1)} - 9.51}{9.51^{(3+1)} - 1} \\ &= 0.99896 \end{aligned}$$

For C_8H_{10}, the fraction stripped is

$$\begin{aligned} f_{C_8H_{10}} &= \frac{S_{C_8H_{10}}^{N+1} - S_{C_8H_{10}}}{S_{C_8H_{10}}^{N+1} - 1} \\ &= \frac{10.93^{(3+1)} - 10.93}{10.93^{(3+1)} - 1} \\ &= 0.99930 \end{aligned}$$

After stripping, the concentration of C_6H_6 in the wastewater stream is

$$\begin{aligned} C_{C_6H_6,\text{out}} &= (1 - f_{C_6H_6})C_{C_6H_6} \\ &= (1 - 0.99905)\left(160\ \frac{\text{mg}}{\text{L}}\right) \\ &= 0.152\ \text{mg/L} \end{aligned}$$

After stripping, the concentration of C_7H_8 in the wastewater stream is

$$\begin{aligned} C_{C_7H_8,\text{out}} &= (1 - f_{C_7H_8})C_{C_7H_8} \\ &= (1 - 0.99896)\left(45\ \frac{\text{mg}}{\text{L}}\right) \\ &= 0.0468\ \text{mg/L} \end{aligned}$$

After stripping, the concentration of C_8H_{10} in the wastewater stream is

$$\begin{aligned} C_{C_8H_{10},\text{out}} &= (1 - f_{C_8H_{10}})C_{C_8H_{10}} \\ &= (1 - 0.99930)\left(20\ \frac{\text{mg}}{\text{L}}\right) \\ &= 0.0140\ \text{mg/L} \end{aligned}$$

The total VOCs remaining in the wastewater is

$$\begin{aligned} C_{\text{VOC}} &= C_{C_6H_6,\text{out}} + C_{C_7H_8,\text{out}} + C_{C_8H_{10},\text{out}} \\ &= 0.152\ \frac{\text{mg}}{\text{L}} + 0.0468\ \frac{\text{mg}}{\text{L}} + 0.0140\ \frac{\text{mg}}{\text{L}} \\ &= 0.2128\ \text{mg/L} \quad (0.21\ \text{mg/L}) \end{aligned}$$

The answer is (B).

31. From *NCEES Handbook* table "Dimensionless Numbers," the average Reynolds number is

$$\begin{aligned} Re_w &= \frac{\rho u D_{\text{inner}} \text{v}\rho}{\mu_w} = \frac{D_{\text{inner}}\dot{m}_w}{A\mu_w} = \frac{D_{\text{inner}}\dot{m}_w}{\left(\frac{\pi D_{\text{inner}}^2}{4}\right)\mu_w} \\ &= \frac{4\dot{m}_w}{\pi D_{\text{inner}}\mu_w} \\ &= \frac{(4)\left(0.2\ \frac{\text{kg}}{\text{s}}\right)}{\pi(0.025\ \text{m})\left(725 \times 10^{-6}\ \frac{\text{N}\cdot\text{s}}{\text{m}^2}\right)} \\ &= 14\,050 \end{aligned}$$

The Reynolds number for the water side of the heat exchanger is larger than 2000, so the flow through the inner pipe is turbulent. The Dittus-Boelter equation is applicable to the turbulent flow of the water side and over these ranges of dimensionless groups: $0.7 \le \text{Pr}_w \le 160$ and $\text{Re}_w \ge 10\,000$. The average Nusselt number for the water side, from the Dittus-Boelter equation, is

$$\begin{aligned} \overline{Nu}_w &= 0.023(Re_w)^{4/5}(Pr_w)^{0.4} \\ &= (0.023)(14\,050)^{4/5}(4.85)^{0.4} \\ &= 89.98 \end{aligned}$$

From *NCEES Handbook* table "Forced Convection—External Flow," the average Nusselt number is

$$\overline{Nu} = \frac{\bar{h}D}{k}$$

Solving the preceding equation for the average convection coefficient, h, for the water side of the heat exchanger

$$\bar{h}_w = (\overline{Nu}_w)\left(\frac{k_w}{D_{\text{inner}}}\right) = (89.98)\left(\frac{0.625\ \frac{\text{W}}{\text{m}\cdot\text{K}}}{0.025\ \text{m}}\right) = 2249.50\ \text{W/m}^2\cdot\text{K}$$

From *NCEES Handbook* table "Forced Convection—External Flow," the hydraulic diameter of the annulus is

$$D_{\text{H}} = D_{\text{outer}} - D_{\text{inner}} = 0.045\ \text{m} - 0.025\ \text{m} = 0.020\ \text{m}$$

Use the hydraulic diameter to find the Reynolds number for the hydrocarbon side of the heat exchanger.

$$Re_h = \frac{\rho u D_{\text{H}}}{\mu_h} = \frac{D_{\text{H}}\dot{m}_h}{A\mu_h} = \frac{D_{\text{H}}\dot{m}_h}{\left(\frac{\pi D_{\text{H}}^2}{4}\right)\mu_h} = \frac{4\dot{m}_h}{\pi D_{\text{H}}\mu_h} = \frac{(4)\left(0.1\ \frac{\text{kg}}{\text{s}}\right)}{\pi(0.020\ \text{m})\left(0.0325\ \frac{\text{N}\cdot\text{s}}{\text{m}^2}\right)} = 195.88$$

The Reynolds number for the hydrocarbon side is less than 2000, so the annular flow is laminar. The average Nusselt number for the hydrocarbon side is given as 5.56. The convection coefficient on this side is

$$\bar{h}_h = \frac{k_h(\overline{Nu}_h)}{D_{\text{H}}} = \frac{\left(0.138\ \frac{\text{W}}{\text{m}\cdot\text{K}}\right)(5.56)}{0.020\ \text{m}} = 38.36\ \text{W/m}^2\cdot\text{K}$$

From *NCEES Handbook:* Overall Heat-Transfer Coefficient, the overall heat transfer coefficient is

$$\frac{1}{U_{\text{ov}}A_{\text{ref}}} = \frac{1}{h_{\text{i}}A_{\text{i}}} + \frac{R_{\text{fi}}}{A_{\text{i}}} + \frac{\ln\left(\frac{D_o}{D_i}\right)}{2\pi kL} + \frac{R_{\text{fo}}}{A_{\text{o}}} + \frac{1}{h_{\text{o}}A_{\text{o}}}$$

The thermal resistance of the pipe walls, heat loss to the surroundings, and fouling factors are all negligible, so the overall convection coefficient is

$$U_{\text{ov}} = \frac{1}{\frac{1}{\bar{h}_w} + \frac{1}{\bar{h}_h}} = \frac{1}{\frac{1}{2249.50\ \frac{\text{W}}{\text{m}^2\cdot\text{K}}} + \frac{1}{38.36\ \frac{\text{W}}{\text{m}^2\cdot\text{K}}}} = 37.72\ \text{W/m}^2\cdot\text{K}$$

The required heat transfer rate is

$$\dot{Q} = \dot{m}_w c_{\text{p},w}(T_{\text{cold,out}} - T_{\text{cold,in}}) = \left(0.2\ \frac{\text{kg}}{\text{s}}\right)\left(4178\ \frac{\text{J}}{\text{kg}\cdot{}^\circ\text{C}}\right)(80^\circ\text{C} - 60^\circ\text{C}) = 16\,712\ \text{W}$$

From *NCEES Handbook:* Log-Mean Temperature Difference, the logarithmic mean temperature difference is

$$\Delta T_{\text{lm}} = \frac{(T_{\text{hot,out}} - T_{\text{cold,in}}) - (T_{\text{hot,in}} - T_{\text{cold,out}})}{\ln\left(\frac{T_{\text{hot,out}} - T_{\text{cold,in}}}{T_{\text{hot,in}} - T_{\text{cold,out}}}\right)} = \frac{(100^\circ\text{C} - 60^\circ\text{C}) - (150^\circ\text{C} - 80^\circ\text{C})}{\ln\frac{100^\circ\text{C} - 60^\circ\text{C}}{150^\circ\text{C} - 80^\circ\text{C}}} = 53.61^\circ\text{C}$$

From *NCEES Handbook:* Combination of Heat-Transfer Mechanisms, the length of the inner pipe is found from the equation

$$\dot{Q} = U_{\text{ov}}A\Delta T = UA\Delta T_{\text{lm}}$$

The area in this case is the heating area, or superficial area, which is

$$A = \pi L D_{\text{inner}}$$

So,

$$\dot{Q} = U_{ov}(\pi L D_{inner})\Delta T_{lm}$$

$$L = \frac{\dot{Q}}{U_{ov}\pi D_{inner}\Delta T_{lm}}$$

$$= \frac{16\,712 \text{ W}}{\left(37.72 \ \frac{\text{W}}{\text{m}^2\cdot\text{K}}\right)\pi(0.025 \text{ m})(53.61°\text{C})}$$

$$= 105.22 \text{ m} \quad (110 \text{ m})$$

The answer is (C).

32. Calculate based on a unit surface area, A, of 1 m^2. The volumetric flow rate of the water, $\dot{V}_s$, is 15 m^3/d. From *NCEES Handbook:* Pneumatic Transport, the superficial solids velocity is

$$\bar{u}_s = \frac{\dot{V}_s}{A}$$

The average overflow velocity is

$$\bar{u}_s = \frac{\dot{V}_s}{A} = \frac{\left(15 \ \frac{\text{m}^3}{\text{d}}\right)\left(1000 \ \frac{\text{mm}}{\text{m}}\right)}{(1 \text{ m}^2)\left(86\,400 \ \frac{\text{s}}{\text{d}}\right)} = 0.1736 \text{ mm/s}$$

The particles being considered have a settling velocity, $\bar{u}_{set}$, of 0.1 mm/s. This is less than the average overflow velocity, so some fraction of these particles will overflow into the water treatment tank. The percentage of particles with a settling velocity of 0.1 mm/s that will settle under these conditions is

$$P = \frac{\bar{u}_{set}}{\bar{u}_s} = \frac{0.1 \ \frac{\text{mm}}{\text{s}}}{0.1736 \ \frac{\text{mm}}{\text{s}}} \times 100\% = 57.6\% \quad (60\%)$$

The answer is (C).

33. The symbols matched to the correct valve names are shown.

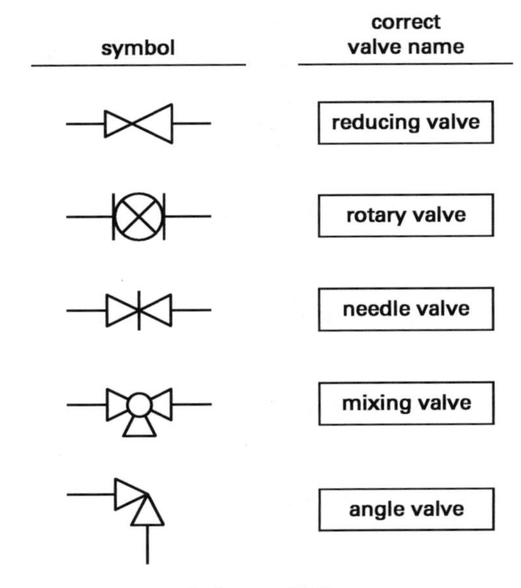

The correct answer choices are shown.

34. The steam enters at 256°F, and the vapor produced is at 220°F, so the temperature difference in the evaporator is

$$\Delta T = T_{steam} - T_{vapor}$$
$$= 256°\text{F} - 220°\text{F}$$
$$= 36°\text{F}$$

The area of heat transfer, A, is given as 150 ft^2. The rate of heat transfer is

$$\dot{Q} = UA\Delta T$$

$$= \frac{\left(785 \ \frac{\text{Btu}}{\text{hr-ft}^2\text{-°F}}\right)(150 \text{ ft}^2)(36°\text{F})}{3600 \ \frac{\text{sec}}{\text{hr}}}$$

$$= 1177.5 \text{ Btu/sec}$$

From *NCEES Handbook* table "Saturated Steam (U.S. Units)—Temperature Table," the latent heat of the vapor at 220°F, λ_{vapor}, is 1153.4 Btu/lbm. An energy balance around the evaporator is

$$\dot{m}_{vapor}\lambda_{vapor} = \dot{Q}$$

So, the mass flow rate of water evaporated is

$$\dot{m}_{vapor} = \frac{\dot{Q}}{\lambda_{vapor}} = \frac{1177.5 \ \frac{\text{Btu}}{\text{sec}}}{1153.4 \ \frac{\text{Btu}}{\text{lbm}}} = 1.0209 \text{ lbm/sec}$$

The feed is 6% solute and the solution from the evaporator is 15% solute. So, the mass balance of solute around the evaporator is

$$0.06m_{in} = 0.15m_{out}$$

m_{in} and m_{out} are the total masses of the feed and the outgoing solution, respectively. Solving for m_{out} gives

$$m_{out} = \frac{0.06m_{in}}{0.15} = \frac{(0.06)(40{,}000 \text{ lbm})}{0.15} = 16{,}000 \text{ lbm}$$

The feed is 6% starch, so it is 94% water. The water in the feed is

$$m_{water,in} = 0.94m_{in} = (0.94)(40{,}000 \text{ lbm}) = 37{,}600 \text{ lbm}$$

Similarly, the solution produced in the evaporator is 15% starch, so it is also 85% water. The water in the solution is

$$m_{water,out} = 0.85m_{out} = (0.85)(16{,}000 \text{ lbm}) = 13{,}600 \text{ lbm}$$

From *NCEES Handbook:* Material Balances, the general balance equation is

$$Accumulation = Input - Output + Generation - Consumption$$

Use a water mass balance around the evaporator to find the water evaporated.

$$m_{vapor} = m_{water,in} - m_{water,out} = 37{,}600 \text{ lbm} - 13{,}600 \text{ lbm} = 24{,}000 \text{ lbm}$$

The time it takes for the water to evaporate, and therefore for the starch solution to reach 15% concentration, is

$$t = \frac{m_{vapor}}{\dot{m}_{vapor}} = \frac{24{,}000 \text{ lbm}}{\left(1.0209 \ \frac{\text{lbm}}{\text{sec}}\right)\left(3600 \ \frac{\text{sec}}{\text{hr}}\right)} = 6.53019 \text{ hr} \quad (6.5 \text{ hr})$$

The answer is (C).

35. For the irreversible reaction, the rate law is given as

$$-r_M = k_M C_M C_N^{0.5}$$

The rate law for the reversible reaction will reduce to the rate law of the irreversible reaction when the concentration of the species R reduces to zero. Start with option A, replacing C_R with zero.

The equation in option A is

$$-r_M = k_M\left(C_M C_N^{0.5} - \frac{C_R}{\sqrt{K}}\right)$$

Replacing C_R with zero gives

$$-r_M = k_M\left(C_M C_N^{0.5} - \frac{0}{\sqrt{K}}\right) = k_M C_M C_N^{0.5}$$

This equation equals the rate law given in the problem statement. Among the four options, only A and B reduce to the rate law when making the concentration of the species R equal to zero.

For the reversible reaction, the equilibrium concentration constant is

$$K = \frac{C_R^2}{C_M^2 C_N}$$

Taking the square root gives

$$\sqrt{K} = \frac{C_R}{C_M C_N^{0.5}}$$

Rearranging,

$$C_M C_N^{0.5} = \frac{C_R}{\sqrt{K}}$$

$$C_M C_N^{0.5} - \frac{C_R}{\sqrt{K}} = 0$$

Substituting this equation in option A gives

$$-r_M = k_M(0)$$

This makes the reaction rate at equilibrium zero, and there is no need to check the other options.

The answer is (A).

36. The relevant data are shown in the illustration.

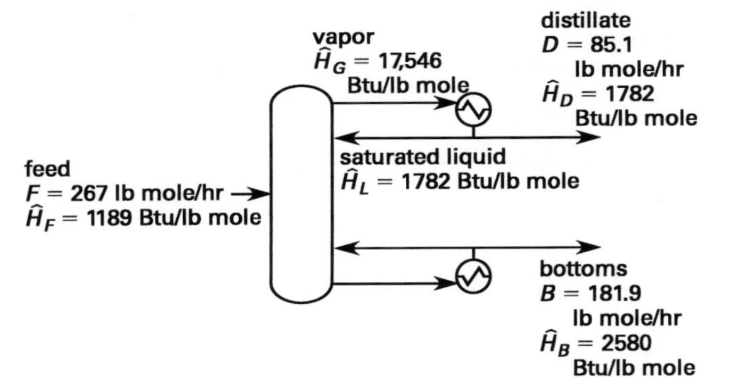

The reflux ratio, R, is given as 1.038. The *NCEES Handbook:* Material and Energy Balances for Trayed and Packed Units, the overall, component and energy balances are used extensively throughout the solution of this problem. To perform the energy balance around the top of the distillation column, let the basis of calculation be 1 lb mole/hr of distillate. The heat removed in the condenser and the permanently removed distillate is

$$\begin{aligned}\widehat{H}'_D &= R\left(\widehat{H}_G - \widehat{H}_L\right) + \widehat{H}_G \\ &= (1.038)\left(17{,}546\ \frac{\text{Btu}}{\text{lb mole}} - 1782\ \frac{\text{Btu}}{\text{lb mole}}\right) \\ &\quad +17{,}546\ \frac{\text{Btu}}{\text{lb mole}} \\ &= 33{,}909\ \text{Btu/lb mole}\end{aligned}$$

The total heat removed in the reboiler and the permanently removed bottoms stream is

$$F\widehat{H}_F = D\widehat{H}'_D + B\widehat{H}'_B$$

Solving for the enthalpy of the bottoms stream gives

$$\begin{aligned}\widehat{H}'_B &= \frac{F\widehat{H}_F - D\widehat{H}'_D}{B} \\ &= \frac{\left(\begin{array}{l}\left(267\ \frac{\text{lb mole}}{\text{hr}}\right)\left(1189\ \frac{\text{Btu}}{\text{lb mole}}\right) - \left(85.1\ \frac{\text{lb mole}}{\text{hr}}\right) \\ \times\left(33{,}909\ \frac{\text{Btu}}{\text{lb mole}}\right)\end{array}\right)}{181.9\ \frac{\text{lb mole}}{\text{hr}}} \\ &= -14{,}119\ \text{Btu/lb mole}\end{aligned}$$

An energy balance around the reboiler gives the heat load in the reboiler.

$$\widehat{H}'_B = \widehat{H}_B - \frac{Q_B}{B}$$

Solving for the heat load in the reboiler gives

$$\begin{aligned}Q_B &= B\widehat{H}_B - B\widehat{H}'_B \\ &= \left(181.9\ \frac{\text{lb mole}}{\text{hr}}\right)\left(2580\ \frac{\text{Btu}}{\text{lb mole}}\right) - \left(181.9\ \frac{\text{lb mole}}{\text{hr}}\right) \\ &\quad \times\left(-14{,}118\ \frac{\text{Btu}}{\text{lb mole}}\right) \\ &= 3{,}037{,}000\ \text{Btu/hr} \quad (3.0\times 10^6\ \text{Btu/hr})\end{aligned}$$

The answer is (D).

37. The manometer reading, R, is given as 1.65 in Hg, and the specific gravity of mercury, SG_{Hg}, is given as 13.6. The height of the column of water is

$$h = R(SG_{Hg}) = \frac{(1.65\ \text{in Hg})\left(13.6\ \frac{\text{in wg}}{\text{in Hg}}\right)}{12\ \frac{\text{in}}{\text{ft}}} = 1.87\ \text{ft}$$

From *NCEES Handbook:* Orifice Discharging Freely into Atmosphere, the velocity of the water at the center of the pipe is

$$u = \sqrt{2gh} = \sqrt{(2)\left(32.2\ \frac{\text{ft}}{\text{sec}^2}\right)(1.87\ \text{ft})} = 10.97\ \text{ft/sec}$$

The average velocity is 96.9% of the velocity at the center of the pipe.

$$u_{avg} = 0.969u = (0.969)\left(10.97\ \frac{\text{ft}}{\text{sec}}\right) = 10.63\ \text{ft/sec}$$

The internal cross-sectional area of the pipe is

$$A = \frac{\pi d^2}{4} = \frac{\pi\left(\dfrac{2.469\ \text{in}}{12\ \dfrac{\text{in}}{\text{ft}}}\right)^2}{4} = 0.03325\ \text{ft}^2$$

The density of the water, ρ, is given as 62.4 lbm/ft^3. From *NCEES Handbook:* Conservation of Mass, the continuity equation is

$$\rho_1 A_1 u_1 = \rho_2 A_2 u_2$$

Applying the preceding equation to the conditions of the problem statement, the mass flow rate of the water is

$$\begin{aligned} \dot{m} &= \rho A u_{\text{avg}} \\ &= \left(62.4\ \frac{\text{lbm}}{\text{ft}^3}\right)(0.03325\ \text{ft}^2)\left(\left(10.63\ \frac{\text{ft}}{\text{sec}}\right)\left(60\ \frac{\text{sec}}{\text{min}}\right)\right) \\ &= 1323\ \text{lbm/min} \quad (1300\ \text{lbm/min}) \end{aligned}$$

The answer is (A).

38. The feed contains 1.5% solids, and the product contains 25% solids. A solid mass balance around the evaporator, then, gives

$$0.015\dot{m}_F = 0.25\dot{m}_L$$

Solving for $\dot{m}_L$,

$$\dot{m}_L = \frac{0.015\dot{m}_F}{0.25} = \frac{(0.015)\left(15\,000\ \dfrac{\text{kg}}{\text{h}}\right)}{0.25} = 900\ \text{kg/h}$$

A total mass balance around the evaporator gives the mass flow rate of water vaporized.

$$\begin{aligned} \dot{m}_v &= \dot{m}_F - \dot{m}_L \\ &= 15\,000\ \frac{\text{kg}}{\text{h}} - 900\ \frac{\text{kg}}{\text{h}} \\ &= 14\,100\ \text{kg/h} \end{aligned}$$

The evaporator is at a pressure of 22.88 kPa. From *NCEES Handbook* table "Saturated Steam (SI Units)—Temperature Table," the normal boiling point of the water in the evaporator, T_b, is 63°C. The boiling point elevation (BPE) is given as 21°C. The temperature of the water vapor leaving the evaporator is

$$T_v = \text{BPE} + T_b = 63°\text{C} + 21°\text{C} = 84°\text{C}$$

This is also the temperature of the solution produced in the evaporator. The enthalpies of the feed and the product are assumed to equal the enthalpy of pure water at the same temperatures. The feed is at 25°C, so its specific enthalpy, h_F, can be found in *NCEES Handbook* table "Saturated Steam (SI Units)—Temperature Table," and equals 104.89 kJ/kg. The solution produced in the evaporator is at 84°C, so the steam table gives its specific enthalpy, h_L, as 351.702 kJ/kg. The vapor produced in the evaporator is at 84°C, and the steam table gives its specific enthalpy, h_v, as 2650.26 kJ/kg. The saturated water is at 22.88 kPa, and the steam table gives its specific enthalpy, h_w, as 263.23 kJ/kg. The steam enters the evaporator at 120°C, so from the steam table, the latent heat of the steam, λ, is 2683.42 kJ/kg. An energy balance around the evaporator gives

$$\dot{m}_F h_F + \dot{m}_s \lambda = \dot{m}_v h_v + \dot{m}_L h_L$$

Solving for the steam flow rate gives

$$\begin{aligned} \dot{m}_s &= \frac{\dot{m}_v h_v + \dot{m}_L h_L - \dot{m}_F h_F}{\lambda} \\ &= \frac{\left(\begin{array}{c} \left(14\,100\ \dfrac{\text{kg}}{\text{h}}\right)\left(2650.26\ \dfrac{\text{kJ}}{\text{kg}}\right) \\ +\left(900\ \dfrac{\text{kg}}{\text{h}}\right)\left(351.702\ \dfrac{\text{kJ}}{\text{kg}}\right) \\ -\left(15\,000\ \dfrac{\text{kg}}{\text{h}}\right)\left(104.89\ \dfrac{\text{kJ}}{\text{kg}}\right) \end{array}\right)}{\left(2683.42\ \dfrac{\text{kJ}}{\text{kg}}\right)\left(3600\ \dfrac{\text{s}}{\text{h}}\right)} \\ &= 3.74\ \text{kg/s} \end{aligned}$$

The overall heat transfer coefficient, U, is given as 2750 W/m^2·°C. The temperature difference is

$$\begin{aligned} \Delta T &= T_s - T_w \\ &= 120°\text{C} - 84°\text{C} \\ &= 36°\text{C} \end{aligned}$$

The rate of heat transfer from the steam is

$$\dot{m}_s \lambda = UA\Delta T$$

Solving for the heat transfer area gives

$$A = \frac{\dot{m}_s \lambda}{U \Delta T} = \frac{\left(3.74\ \frac{\text{kg}}{\text{s}}\right)\left(2683.42\ \frac{\text{kJ}}{\text{kg}}\right)\left(1000\ \frac{\text{J}}{\text{kJ}}\right)}{\left(2750\ \frac{\text{W}}{\text{m}^2\cdot{}^\circ\text{C}}\right)(36^\circ\text{C})} = 101.4\ \text{m}^2 \quad (100\ \text{m}^2)$$

The answer is (C).

39. The relevant data are shown in the illustration.

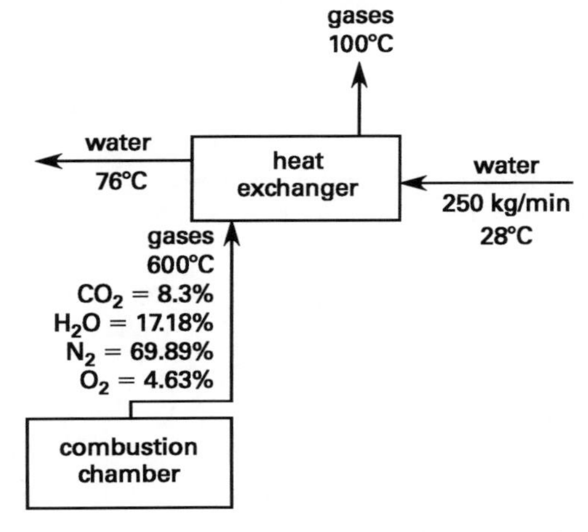

The mass flow rate of the water in the heat exchanger is given as

$$\dot{m} = \left(250\ \frac{\text{kg}}{\text{min}}\right)\left(1000\ \frac{\text{g}}{\text{kg}}\right) = 250\,000\ \text{g/min}$$

The heat capacity of the water, c_p, is given as 4.184 J/g·°C. From the *NCEES Handbook:* Heat Capacity/Specific Heat (c_p), the heat transferred is

$$\dot{Q} = \dot{m} c_p \Delta T$$

The enthalpy change of the water is

$$\dot{Q}_{\text{water}} = \dot{m} c_p (T_{\text{water,out}} - T_{\text{water,in}}) = \left(250\,000\ \frac{\text{g}}{\text{min}}\right)\left(4.184\ \frac{\text{J}}{\text{g}\cdot{}^\circ\text{C}}\right)(76^\circ\text{C} - 28^\circ\text{C}) = 50\,208\,000\ \text{J/min}$$

The feed is 8.3% CO_2, 17.18% H_2O, 69.89% N_2, and the remainder O_2, so the number of moles of each component in 1 kmol of gases from the combustion chamber is

$$n_{CO_2} = 0.083\ \text{mol}$$
$$n_{H_2O} = 0.1718\ \text{mol}$$
$$n_{N_2} = 0.6989\ \text{mol}$$
$$n_{O_2} = n_{\text{total}} - n_{CO_2} - n_{H_2O} - n_{N_2} = 1.0\ \text{mol} - 0.083\ \text{mol} - 0.1718\ \text{mol} - 0.6989\ \text{mol} = 0.0463\ \text{mol}$$

Base the calculations on 1 mol of combustion gases. From the table of enthalpies in the problem statement, the difference in enthalpy in the CO_2 between 600°C and 100°C is

$$\Delta H_{CO_2} = n_{CO_2}(\widehat{H}_{CO_2,600^\circ C} - \widehat{H}_{CO_2,100^\circ C}) = (0.083\ \text{mol})\left(26.53\ \frac{\text{kJ}}{\text{mol}} - 2.90\ \frac{\text{kJ}}{\text{mol}}\right) = 1.961\ \text{kJ}$$

The difference in enthalpy in the H_2O between 600°C and 100°C is

$$\Delta H_{H_2O} = n_{H_2O}(\widehat{H}_{H_2O,600^\circ C} - \widehat{H}_{H_2O,100^\circ C}) = (0.1718\ \text{mol})\left(20.91\ \frac{\text{kJ}}{\text{mol}} - 2.54\ \frac{\text{kJ}}{\text{mol}}\right) = 3.156\ \text{kJ}$$

The difference in enthalpy in the N_2 between 600°C and 100°C is

$$\Delta H_{N_2} = n_{N_2}(\widehat{H}_{N_2,600^\circ C} - \widehat{H}_{N_2,100^\circ C}) = (0.6989\ \text{mol})\left(17.39\ \frac{\text{kJ}}{\text{mol}} - 2.19\ \frac{\text{kJ}}{\text{mol}}\right) = 10.62\ \text{kJ}$$

The difference in enthalpy in the O_2 between 600°C and 100°C is

$$\Delta H_{O_2} = n_{O_2}(\widehat{H}_{O_2,600^\circ C} - \widehat{H}_{O_2,100^\circ C}) = (0.0463\ \text{mol})\left(18.41\ \frac{\text{kJ}}{\text{mol}} - 2.24\ \frac{\text{kJ}}{\text{mol}}\right) = 0.749\ \text{kJ}$$

The total enthalpy change of 1 mol of combustion gases in the heat exchanger is

$$\Delta H_{\text{gases}} = \Delta H_{CO_2} + \Delta H_{H_2O} + \Delta H_{N_2} + \Delta H_{O_2} = 1.961\ \text{kJ} + 3.156\ \text{kJ} + 10.62\ \text{kJ} + 0.749\ \text{kJ} = 16.486\ \text{kJ} \quad \left(\begin{array}{c}16\,486\ \text{J per 1 mol of} \\ \text{combustion gases}\end{array}\right)$$

An energy balance around the heat exchanger gives the molar flow rate of the combustion gases entering the heat exchanger.

$$\begin{aligned} n &= \frac{\dot{Q}_{\text{water}}}{\Delta H_{\text{gases}}} \\ &= \frac{50\,208\,000 \ \frac{\text{J}}{\text{min}}}{16\,486 \ \frac{\text{J}}{\text{mol}}} \\ &= 3045.5 \ \text{mol/min} \quad (3.0 \times 10^3 \ \text{mol/min}) \end{aligned}$$

The answer is (B).

40. From *NCEES Handbook:* Material Balances, the general balance equation

$$\begin{aligned} \textit{Accumulation} &= \textit{Input} - \textit{Output} \\ &\quad + \textit{Generation} - \textit{Consumption} \end{aligned}$$

is used extensively throughout this problem solution.

The feed entering the crystallizer is composed of water and ammonium sulfate.

$$m_{F,\text{total}} = m_{F,\text{water}} + m_{F,\text{sulfate}} = 4800 \ \text{lbm}$$

The sulfate makes up 25% of the feed by weight.

$$\begin{aligned} m_{F,\text{sulfate}} &= 0.25 m_{F,\text{total}} \\ &= (0.25)(4800 \ \text{lbm}) \\ &= 1200 \ \text{lbm} \\ m_{F,\text{water}} &= m_{F,\text{total}} - m_{F,\text{sulfate}} \\ &= 4800 \ \text{lbm} - 1200 \ \text{lbm} \\ &= 3600 \ \text{lbm} \end{aligned}$$

75% of the water is extracted in the form of vapor.

$$\begin{aligned} m_{\text{vapor}} &= 0.75 m_{F,\text{water}} \\ &= (0.75)(3600 \ \text{lbm}) \\ &= 2700 \ \text{lbm} \end{aligned}$$

The amount of water remaining in the crystallizer after evaporation is found by a water mass balance around the crystallizer.

$$\begin{aligned} m_{r,\text{water}} &= m_{F,\text{water}} - m_{\text{vapor}} \\ &= 3600 \ \text{lbm} - 2700 \ \text{lbm} \\ &= 900 \ \text{lbm} \end{aligned}$$

The sulfate from the feed also remains in the crystallizer.

$$\begin{aligned} m_{r,\text{sulfate}} &= m_{F,\text{sulfate}} \\ &= 1200 \ \text{lbm} \end{aligned}$$

Some of this sulfate and water crystallizes. Because the stable solid phase is the monohydrate, the ratio of the water of crystallization to the crystallized sulfate is equal to the ratio of their molecular weights.

$$\begin{aligned} \frac{m_{c,\text{water}}}{m_{c,\text{sulfate}}} &= \frac{MW_{\text{water}}}{MW_{\text{sulfate}}} \\ &= \frac{18 \ \frac{\text{lbm}}{\text{lb mole}}}{132 \ \frac{\text{lbm}}{\text{lb mole}}} \\ m_{c,\text{water}} &= 0.136 m_{c,\text{sulfate}} \end{aligned}$$

The rest of the sulfate and water remains in the crystallizer as a solution. A sulfate mass balance gives

$$\begin{aligned} m_{r,\text{sulfate}} &= m_{c,\text{sulfate}} + m_{s,\text{sulfate}} \\ m_{c,\text{sulfate}} &= m_{r,\text{sulfate}} - m_{s,\text{sulfate}} \end{aligned}$$

The solution in the crystallizer is made entirely of water and sulfate.

$$m_{s,\text{total}} = m_{s,\text{water}} + m_{s,\text{sulfate}}$$

A water mass balance around the crystallization process gives

$$m_{s,\text{water}} = m_{r,\text{water}} - m_{c,\text{water}}$$

The solubility of the sulfate is 46 wt%, so

$$\begin{aligned} 0.46 &= \frac{m_{s,\text{sulfate}}}{m_{s,\text{total}}} \\ &= \frac{m_{s,\text{sulfate}}}{m_{s,\text{water}} + m_{s,\text{sulfate}}} \\ &= \frac{m_{s,\text{sulfate}}}{(m_{r,\text{water}} - m_{c,\text{water}}) + m_{s,\text{sulfate}}} \\ &= \frac{m_{s,\text{sulfate}}}{m_{r,\text{water}} - 0.136 m_{c,\text{sulfate}} + m_{s,\text{sulfate}}} \\ &= \frac{m_{s,\text{sulfate}}}{m_{r,\text{water}} - 0.136(m_{r,\text{sulfate}} - m_{s,\text{sulfate}}) + m_{s,\text{sulfate}}} \\ &= \frac{m_{s,\text{sulfate}}}{900 \ \text{lbm} - (0.136)\begin{pmatrix} 1200 \ \text{lbm} \\ - m_{s,\text{sulfate}} \end{pmatrix} + m_{s,\text{sulfate}}} \\ &= \frac{m_{s,\text{sulfate}}}{1.136 m_{s,\text{sulfate}} + 736.8 \ \text{lbm}} \end{aligned}$$

Solving for the amount of sulfate in the solution gives

$$\begin{aligned} m_{s,\text{sulfate}} &= (0.46)(1.136 m_{s,\text{sulfate}} + 736.8 \text{ lbm}) \\ &= 0.523 m_{s,\text{sulfate}} + 338.9 \text{ lbm} \\ 0.477 m_{s,\text{sulfate}} &= 338.9 \text{ lbm} \\ m_{s,\text{sulfate}} &= \frac{338.9 \text{ lbm}}{0.477} \\ &= 710.5 \text{ lbm} \end{aligned}$$

Find the amounts of sulfate and water in the crystals.

$$\begin{aligned} m_{c,\text{sulfate}} &= m_{r,\text{sulfate}} - m_{s,\text{sulfate}} \\ &= 1200 \text{ lbm} - 710.5 \text{ lbm} \\ &= 489.5 \text{ lbm} \\ m_{c,\text{water}} &= 0.136 m_{c,\text{sulfate}} \\ &= (0.136)(489.5 \text{ lbm}) \\ &= 66.57 \text{ lbm} \end{aligned}$$

The total mass of the crystals is

$$\begin{aligned} m_{c,\text{total}} &= m_{c,\text{water}} + m_{c,\text{sulfate}} \\ &= 66.57 \text{ lbm} + 489.5 \text{ lbm} \\ &= 556.07 \text{ lbm} \quad (560 \text{ lbm}) \end{aligned}$$

The answer is (C).

Solutions
Afternoon Session

41. The density of the NaCl solution, ρ, is given as a constant 62.4 lbm/ft^3. The solution of NaCl in the tank is 12%, so the mass fraction of NaCl in the tank, w_{tank}, is 0.12. From *NCEES Handbook* table "Conversion Table for the Most Commonly Used Units of Volume," the volume of 1 ft^3 is 7.4814 gal. The initial mass concentration of the NaCl solution is

$$\begin{aligned}\gamma_0 &= w_{tank}\rho \\ &= (0.12)\left(\frac{62.4\ \frac{lbm}{ft^3}}{7.4814\ \frac{gal}{ft^3}}\right) \\ &= 1.00\ lbm/gal\end{aligned}$$

The mass of the NaCl initially in the tank is

$$\begin{aligned}m_{NaCl,0} &= V_{solution,0}\gamma_0 \\ &= (1000\ gal)\left(1.00\ \frac{lbm}{gal}\right) \\ &= 1000\ lbm\end{aligned}$$

In ten minutes, the volume of solution added to the tank is

$$\begin{aligned}V_{solution,added} &= Qt \\ &= \left(42\ \frac{gal}{min}\right)(10\ min) \\ &= 420\ gal\end{aligned}$$

After ten minutes, then, the volume of solution in the tank is

$$\begin{aligned}V_{solution,total} &= V_{solution,0} + V_{solution,added} \\ &= 1000\ gal + 420\ gal \\ &= 1420\ gal\end{aligned}$$

This is less than the capacity of the tank, 2000 gal, so the tank does not overflow. The solution of NaCl in the feed is 21%, so the mass fraction of NaCl in the feed, w_{added}, is 0.21. The mass concentration of the NaCl solution added to the tank is

$$\begin{aligned}\gamma_{added} &= w_{added}\rho \\ &= (0.21)\left(\frac{62.4\ \frac{lbm}{ft^3}}{7.4814\ \frac{gal}{ft^3}}\right) \\ &= 1.75\ lbm/gal\end{aligned}$$

From *NCEES Handbook:* Concentrations, the mass concentration is

$$\gamma_i = \frac{m_i}{V}$$

Solving the preceding equation for the mass gives

$$m_i = V\gamma_i$$

Applying the preceding equation the mass of the NaCl in the solution added to the tank is

$$\begin{aligned}m_{NaCl,added} &= V_{solution,added}\gamma_{added} \\ &= (420\ gal)\left(1.75\ \frac{lbm}{gal}\right) \\ &= 735\ lbm\end{aligned}$$

After ten minutes, the total mass of the NaCl in the tank is

$$\begin{aligned}m_{NaCl,total} &= m_{NaCl,0} + m_{NaCl,added} \\ &= 1000\ lbm + 735\ lbm \\ &= 1735\ lbm\end{aligned}$$

From *NCEES Handbook:* Concentrations, the mass concentration is

$$\gamma_i = \frac{m_i}{V}$$

Applying the preceding equation to the conditions of the problem statement, after ten minutes, the mass concentration of NaCl in the solution in the tank is

$$\begin{aligned}\gamma_{\text{total}} &= \frac{m_{\text{NaCl,total}}}{V_{\text{solution,total}}} \\ &= \frac{1735 \text{ lbm}}{1420 \text{ gal}} \\ &= 1.222 \text{ lbm/gal} \quad (1.2 \text{ lbm/gal})\end{aligned}$$

The answer is (B).

42. Options A, B, D, and F all meet one or more of the characteristics of the above maximum liquid level pressures used in design. Options C and E do not meet these conditions.

The answers are A, B, D, and F.

43. The diameter of the orifice is given as

$$D_1 = 1.00 \text{ in}$$

The cross-sectional area of the orifice is

$$\begin{aligned}A &= \frac{\pi D_1^2}{4} \\ &= \frac{\pi \left(\frac{1.00 \text{ in}}{12 \frac{\text{in}}{\text{ft}}} \right)^2}{4} \\ &= 0.00545 \text{ ft}^2\end{aligned}$$

The inside diameter of the pipe is given as

$$D_2 = 2.067 \text{ in}$$

The ratio of the diameter of the orifice to the inside diameter of the pipe is

$$\begin{aligned}\beta &= \frac{D_1}{D_2} \\ &= \frac{1.00 \text{ in}}{2.067 \text{ in}} \\ &= 0.484\end{aligned}$$

The ratio of the pressure differential across taps to the pressure of the gas upstream is

$$\begin{aligned}\phi &= \frac{\Delta P}{P_1} \\ &= \frac{3 \frac{\text{lbf}}{\text{in}^2}}{54.7 \frac{\text{lbf}}{\text{in}^2}} \\ &= 0.0548\end{aligned}$$

The ratio of specific heats is given as

$$k = 1.29$$

The Buckingham expansion factor is

$$\begin{aligned}Y &= 1 - \left(\frac{0.41 + 0.35\beta^4}{k} \right) \phi \\ &= 1 - \left(\frac{0.41 + (0.35)(0.484)^4}{1.29} \right) (0.0548) \\ &= 0.982\end{aligned}$$

The absolute temperature of the gas is

$$T = 50°\text{F} + 460° = 510°\text{R}$$

The discharge coefficient is given as

$$C_d = 1.00$$

The molecular weight of CO_2 is given as

$$MW = 44 \text{ lbm/lb mole}$$

From *NCEES Handbook:* Density and Relative Density the density of the gas is

$$\rho = \frac{m}{V}$$

From *NCEES Handbook:* Ideal Gas Law,

$$PV = \frac{mRT}{MW}$$

Solving the preceding equation for the mass over the volume gives

$$\frac{m}{V} = \frac{(MW)P}{RT}$$

Replacing the preceding equation in the density expression gives the density of the gas at upstream conditions is

$$\begin{aligned}\rho &= \frac{m}{V} = \frac{(MW)P_1}{R^*T} \\ &= \frac{\left(44\ \frac{\text{lbm}}{\text{lb mole}}\right)\left(\left(54.7\ \frac{\text{lbf}}{\text{in}^2}\right)\left(12\ \frac{\text{in}}{\text{ft}}\right)^2\right)}{\left(1545\ \frac{\text{ft-lbf}}{\text{lb mole-°R}}\right)(510\text{°R})} \\ &= 0.4398\ \text{lbm/ft}^3\end{aligned}$$

R^* is the universal gas constant. From *NCEES Handbook:* Square-Edge Orifice Meter (Vena Contracta Taps), the velocity of the gas stream through the square-edged orifice is

$$\begin{aligned}u &= \frac{C_d}{\sqrt{1-\left(\frac{D_1}{D_2}\right)^4}}\sqrt{\frac{2g_c\Delta P}{\rho}} \\ &= \frac{C_d}{\sqrt{1-\beta^4}}\sqrt{\frac{2g_c\Delta P}{\rho}} \\ &= \frac{1.00}{\sqrt{1-(0.484)^4}}\sqrt{\frac{(2)\left(32.2\ \frac{\text{ft-lbm}}{\text{lbf-sec}^2}\right)\left(3600\ \frac{\text{sec}}{\text{hr}}\right)^2 \times\left(\left(3\ \frac{\text{lbf}}{\text{in}^2}\right)\left(12\ \frac{\text{in}}{\text{ft}}\right)^2\right)}{0.4398\ \frac{\text{lbm}}{\text{ft}^3}}} \\ &= 931{,}400\ \text{ft/hr}\end{aligned}$$

The mass flow rate of the gas stream is

$$\begin{aligned}\dot{m} &= Au\rho Y \\ &= (0.00545\ \text{ft}^2)\left(931{,}400\ \frac{\text{ft}}{\text{hr}}\right)\left(0.4398\ \frac{\text{lbm}}{\text{ft}^3}\right)(0.982) \\ &= 2192\ \text{lbm/hr}\quad (2200\ \text{lbm/hr})\end{aligned}$$

The answer is (D).

44. Find the feed rate that will produce the maximum allowable fraction of magnesium sulfate in the bottoms from column 2. From *NCEES Handbook:* Material Balances, the balance equation is

$$\begin{aligned}\textit{Accumulation} &= \textit{Input} - \textit{Output} \\ &\quad + \textit{Generation} - \textit{Consumption}\end{aligned}$$

An overall mass balance around the system gives

$$\begin{aligned}F &= D_1 + D_2 + B_2 \\ B_2 &= F - D_1 - D_2 \\ &= F - 1000\ \frac{\text{lbm}}{\text{hr}} - 1200\ \frac{\text{lbm}}{\text{hr}} \\ &= F - 2200\ \frac{\text{lbm}}{\text{hr}}\end{aligned}$$

The fraction of magnesium sulfate in the feed is 0.00675. The maximum allowable fraction in the bottoms from column 2 is 0.015. The mass balance of magnesium sulfate around the system gives

$$0.00675F = 0.015B_2$$

Substituting the value for B_2 in the mass balance equation gives

$$\begin{aligned}0.00675F &= (0.015)\left(F - 2200\ \frac{\text{lbm}}{\text{hr}}\right) \\ F &= \frac{(0.015)\left(2200\ \frac{\text{lbm}}{\text{hr}}\right)}{0.015 - 0.00675} \\ &= 4000\ \text{lbm/hr}\end{aligned}$$

The answer is (C).

45. The volume of the reactor, V, is given as 25 m^3. The feed of M into the reactor, F_{M}, is given as 32 kmol. The concentration of M into the reactor is

$$C_{\text{M},0} = \frac{F_{\text{M}}}{V} = \frac{32\ \text{kmol}}{25\ \text{m}^3} = 1.28\ \text{kmol/m}^3$$

The conversion of M, X, is given as 0.94. From *NCEES Handbook:* Reaction Parameters, the fractional conversion is

$$X_A = \frac{C_{\text{Ao}} - C_{\text{A}}}{C_{\text{Ao}}}$$

Applying the preceding equation to the conditions of the problem statement and solving for the concentration of component M gives the final concentration of M is

$$\begin{aligned}C_{\text{M}} &= C_{\text{M},0}(1 - X) \\ &= \left(1.28\ \frac{\text{kmol}}{\text{m}^3}\right)(1 - 0.94) \\ &= 0.077\ \text{kmol/m}^3\end{aligned}$$

The reaction rate constant, k, is given as 0.96 m^3/kmol·h. The reaction is second order. From *NCEES*

Handbook: Second-Order Reactions, the performance equation for the batch reactor is

$$kt = \frac{1}{C_M} - \frac{1}{C_{M,0}}$$

Solving for t gives

$$t = \frac{\frac{1}{C_M} - \frac{1}{C_{M,0}}}{k} = \frac{\frac{1}{0.077 \frac{\text{kmol}}{\text{m}^3}} - \frac{1}{1.28 \frac{\text{kmol}}{\text{m}^3}}}{0.96 \frac{\text{m}^3}{\text{kmol}\cdot\text{h}}}$$

$$= 12.714 \text{ h} \quad (13 \text{ h})$$

The answer is (D).

46. The feed contains 30% A, so the fraction of component A in the feed, f_A, is 0.30.

The feed contains only A and C, so the fraction of component C in the feed is

$$\begin{aligned} f_C &= 1 - f_A \\ &= 1 - 0.30 \\ &= 0.70 \end{aligned}$$

The following procedure is described in *NCEES Handbook:* Liquid-Liquid Equilibrium for Partially Miscible and Immiscible Systems.

Locate point F on side $\overline{AC}$ of the diagram, such that the percentage of A by weight is 30%. The solvent is pure, so point S on the corner of the triangle represents the solvent. Draw a straight line from point S to point F. This line is the feed line.

The final raffinate is to contain 10% A on a C-free basis, so locate point B on the $\overline{AC}$ side of the triangle, such that the percentage of A by weight is 10%. This point represents the solvent-free final raffinate. Draw a straight line from B to S.

The line $\overline{BS}$ intercepts the equilibrium curve at point B′ on the raffinate side. Point B′ represents the actual raffinate composition leaving the stage.

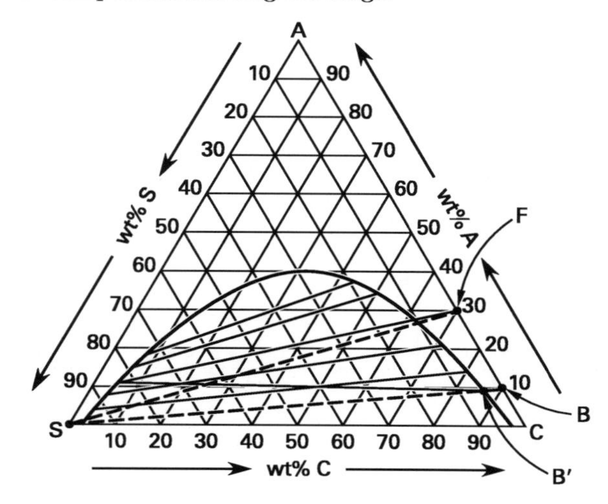

The lines crossing the region with the equilibrium curve are tie lines. Interpolate a new tie line such that if it is extended beyond the right side of the equilibrium curve, it will intersect point F. (This line is very near one of the given tie lines.) This tie line intercepts the equilibrium curve on the extract side at point D′.

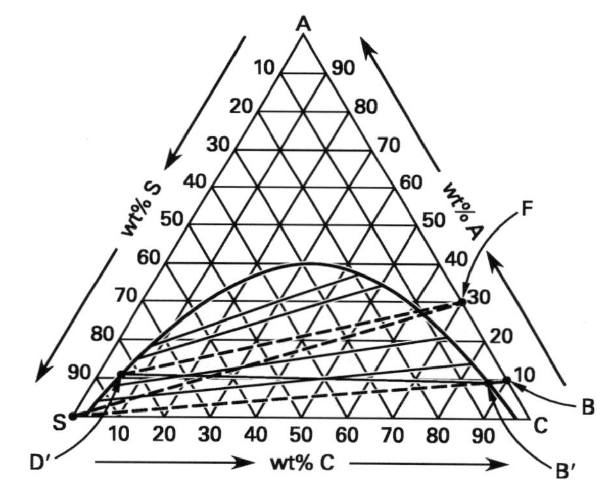

Draw a line from D′ to B′. This line intercepts the feed line at point M. From the diagram, the ratio of the length of the segment $\overline{MF}$ to the segment $\overline{SM}$ is approximately 1.9.

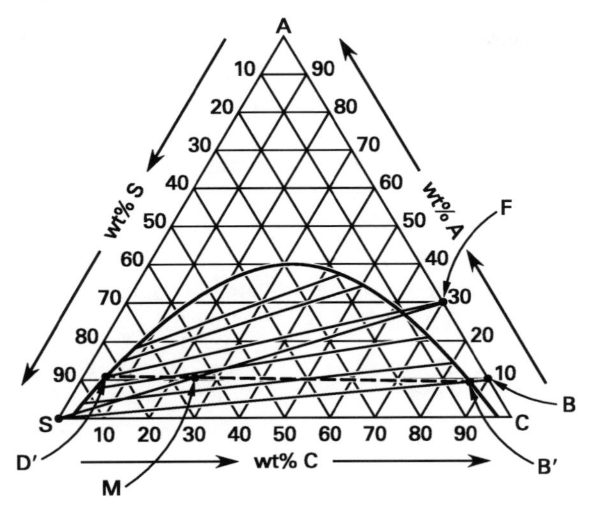

From *NCEES Handbook:* Azeotropes, the lever rule this means that the ratio of the amount of feed to the amount of solvent in this extraction is also approximately 1.9. The amount of solvent used in this extraction is

$$\begin{aligned} S &= 1.9F \\ &= (1.9)(100 \text{ lbm}) \\ &= 190 \text{ lbm} \end{aligned}$$

The answer is (C).

47. From *NCEES Handbook:* Properties of Water, the molar mass of water is 18.01528 lbm/lb mole. From *NCEES Handbook:* Material Balances With Reaction, the molecular weight of dry air is 28.965 lbm/lb mole. The total pressure of the ambient air, P, is given as 14.7 lbf/in^2. From *NCEES Handbook* table "Saturated Steam (U.S. Units)—Temperature Table," the partial pressure of the water in the saturated air at 50°F, P_{w2}, is 0.17796 lbf/in^2. The specific humidity of the saturated air at 50°F is

$$\begin{aligned} \omega_2 &= \left(\frac{P_{w2}}{P-P_{w2}}\right)\left(\frac{MW_{\text{water}}}{MW_{\text{air}}}\right) \\ &= \left(\frac{0.17796\ \frac{\text{lbf}}{\text{in}^2}}{14.7\ \frac{\text{lbf}}{\text{in}^2} - 0.17796\ \frac{\text{lbf}}{\text{in}^2}}\right)\left(\frac{18.01528\ \frac{\text{lbm}}{\text{lb mole}}}{28.965\ \frac{\text{lbm}}{\text{lb mole}}}\right) \\ &= 0.00762 \end{aligned}$$

From *NCEES Handbook* table "Saturated Steam (U.S. Units)—Temperature Table," the specific enthalpy of saturated water vapor at 50°F, h_{wv2}, is found from the steam tables to be 1083.4 Btu/lbm. The makeup water, to be added to the air-water mixture in the adiabatic saturator, is at 50°F, so its specific enthalpy, h_f, is found from the steam tables to be 18.054 Btu/lbm. Also from the steam tables, the specific enthalpy of the water vapor at the adiabatic saturator inlet at 75°F, h_{wv1}, is 1094.3 Btu/lbm. The average heat capacity of the dry air, c_p, is given as 0.24 Btu/lbm-°F. An energy balance around the adiabatic saturator gives

$$\omega_1(h_{wv2}-h_f) - \omega_2(h_{wv1}-h_f) = c_p(T_2-T_1)$$

Solving for the specific humidity at the inlet temperature gives

$$\begin{aligned} \omega_1 &= \frac{c_p(T_2-T_1)+\omega_2(h_{wv1}-h_f)}{h_{wv2}-h_f} \\ &= \frac{\left(\begin{array}{c}\left(0.24\ \frac{\text{Btu}}{\text{lbm-°F}}\right)(50\text{°F}-75\text{°F}) \\ +(0.00762)\left(1094.3\ \frac{\text{Btu}}{\text{lbm}} - 18.054\ \frac{\text{Btu}}{\text{lbm}}\right)\end{array}\right)}{1083.4\ \frac{\text{Btu}}{\text{lbm}} - 18.054\ \frac{\text{Btu}}{\text{lbm}}} \\ &= 0.0021 \end{aligned}$$

The partial vapor pressure in the inlet of the adiabatic saturator is

$$\begin{aligned} P_{w1} &= \left(\frac{\omega_1}{\frac{MW_{\text{water}}}{MW_{\text{air}}}+\omega_1}\right)P \\ &= \left(\frac{0.0021}{\frac{18.01528\ \frac{\text{lbm}}{\text{lb mole}}}{28.965\ \frac{\text{lbm}}{\text{lb mole}}}+0.0021}\right)\left(14.7\ \frac{\text{lbf}}{\text{in}^2}\right) \\ &= 0.0496\ \text{lbf/in}^2 \end{aligned}$$

From *NCEES Handbook* table "Saturated Steam (U.S. Units)—Temperature Table," the partial pressure of the water in the saturated air at 75°F, P_{wv1}, is 0.42964 lbf/in^2. The relative humidity of the ambient air is the ratio of the partial pressure of the water in the inlet of the adiabatic saturator to the partial pressure of the water in the saturated air. Expressed as a percentage, this is

$$\begin{aligned}\phi &= \frac{P_{w1}}{P_{wv1}}\\ &= \frac{0.0496\ \frac{\text{lbf}}{\text{in}^2}}{0.42964\ \frac{\text{lbf}}{\text{in}^2}} \times 100\%\\ &= 11.54\% \quad (12\%)\end{aligned}$$

The answer is (C).

48. The inside diameter of the pipe is

$$\begin{aligned}D &= \frac{10\ \text{cm}}{100\ \frac{\text{cm}}{\text{m}}}\\ &= 0.10\ \text{m}\end{aligned}$$

The cross-sectional area of the pipe is

$$\begin{aligned}A &= \frac{\pi D^2}{4}\\ &= \frac{\pi(0.10\ \text{m})^2}{4}\\ &= 0.007\,85\ \text{m}^2\end{aligned}$$

The volumetric water flow rate is given as

$$\dot{Q} = 0.045\ \text{m}^3/\text{s}$$

The velocity of the water flowing in the pipe is

$$\begin{aligned}u &= \frac{\dot{Q}}{A}\\ &= \frac{0.045\ \frac{\text{m}^3}{\text{s}}}{0.00785\ \text{m}^2}\\ &= 5.73\ \text{m/s}\end{aligned}$$

The specific roughness of the pipe is

$$\begin{aligned}\varepsilon &= \frac{0.047\,\text{mm}}{10^3\ \frac{\text{mm}}{\text{m}}}\\ &= 0.000047\ \text{m}\end{aligned}$$

From *NCEES Handbook:* Absolute Roughness and Relative Roughness, the relative roughness of the pipe is

$$\begin{aligned}\frac{\varepsilon}{D} &= \frac{0.000047\ \text{m}}{0.10\ \text{m}}\\ &= 0.00047\end{aligned}$$

The kinematic viscosity of the water is given as

$$\nu = 1.007 \times 10^{-6}\ \text{m}^2/\text{s}$$

As in *NCEES Handbook* table "Dimensionless Numbers," the Reynolds number is

$$\begin{aligned}Re &= \frac{uvD}{\nu}\\ &= \frac{\left(5.73\ \frac{\text{m}}{\text{s}}\right)(0.10\ \text{m})}{1.007 \times 10^{-6}\ \frac{\text{m}^2}{\text{s}}}\\ &= 5.69 \times 10^5\end{aligned}$$

Given this Reynolds number and specific roughness, the friction factor can be found on the Moody diagram.

$$f = 0.0174$$

From the illustration, the length of the pipe is

$$L = 25\ \text{m} + 20\ \text{m} + 20\ \text{m} = 65\ \text{m}$$

The acceleration due to gravity, g, is 9.80 m/s^2. The fluid encounters a change in cross-sectional area when flowing from the pipe into tank 2, so there is a resistance coefficient to be considered at this point. The resistance coefficient for the entrance, K_{entrance}, is 0.5. The resistance coefficient for the globe valve, K_{valve}, is given as 5.7. The resistance coefficient for the elbow is given as 0.64. There are two elbows in the pipe system, so the resistance coefficient for the elbows is

$$\begin{aligned}K_{\text{elbows}} &= (2)(0.64)\\ &= 1.28\end{aligned}$$

The fluid also encounters a change in cross-sectional area going from tank 1 to the pipe, so there is a resistance coefficient to be considered at this point as well. The resistance coefficient for the exit, K_{exit}, is given as 1.0. Because there are two elbows in the pipe system, from *NCEES Handbook:* Head Loss in Pipe or Conduit, the minor losses are

$$\begin{aligned}K_L &= K_{\text{entrance}} + K_{\text{valve}} + K_{\text{elbows}} + K_{\text{exit}}\\ &= 0.5 + 5.7 + 1.28 + 1.0\\ &= 8.48\end{aligned}$$

Let the elevation at point 0, z_0, be defined as 0 m. From *NCEES Handbook:* Head Loss in Pipe or Conduit, the head loss due to friction is

$$\begin{aligned} h_L &= f\frac{L}{D}\frac{u^2}{2g} + K_L\left(\frac{u^2}{2g}\right) \\ &= (0.0174)\frac{(65\text{ m})}{(0.10\text{ m})}\frac{\left(5.73\ \frac{\text{m}}{\text{s}}\right)^2}{(2)\left(9.80\ \frac{\text{m}}{\text{s}^2}\right)} \\ &\quad + (8.48)\left(\frac{\left(5.73\ \frac{\text{m}}{\text{s}}\right)^2}{(2)\left(9.80\ \frac{\text{m}}{\text{s}^2}\right)}\right) \\ &= 33.15\text{ m} \end{aligned}$$

The elevation of the surface above ground for reservoir 1 is z_1. The elevation of the surface above ground for reservoir 2 is z_2. Each reservoir is open to the atmosphere, so the pressure head is the same for each.

$$P_1 = P_2$$

Because the water flows by gravity from one reservoir to the other, the velocity head is the same for each.

$$u_1 = u_2$$

As in *NCEES Handbook:* The Bernoulli Equation, the Bernoulli equation is

$$\frac{P_1 g_c}{\rho g} + \frac{u_1^2}{2g} + z_1 + h_s = \frac{P_2 g_c}{\rho g} + \frac{u_2^2}{2g} + z_2 + h_L$$

A general mechanical energy balance for flow from point 1 to point 2 with no shaft work, no pump work and with friction is

$$\frac{P_1}{\rho} + z_1 + \frac{u_1^2}{2g} + h_s = \frac{P_2}{\rho} + z_2 + \frac{u_2^2}{2g} + h_L$$

$$\begin{aligned} z_1 - z_2 &= h_L + \left(\frac{P_2}{\rho} - \frac{P_1}{\rho}\right) + \left(\frac{u_2^2}{2g} - \frac{u_1^2}{2g}\right) \\ &= 33.15\text{ m} + 0\text{ m} + 0\text{ m} \\ &= 33.15\text{ m} \quad (33\text{ m}) \end{aligned}$$

The answer is (B).

49. The flow rate is given as

$$\dot{Q} = 660{,}430\text{ gal/day}$$

The reaction constant is given as

$$k = 465\text{ day}^{-1}$$

The concentrations of phenol in the feed and in the product are

$$\begin{aligned} C_{\text{in}} &= 0.00104\text{ lbm/gal} \\ C_{\text{out}} &= 0.000015\text{ lbm/gal} \end{aligned}$$

From *NCEES Handbook:* First-Order Reactions

$$k\tau = \ln\frac{C_{\text{Ao}}}{C_{\text{A}}}$$

Solving for the space time gives

$$\tau = \frac{1}{k}\ln\frac{C_{\text{Ao}}}{C_{\text{A}}} = \frac{V}{\dot{Q}}$$

So, for a plug-flow reactor at steady state with a first-order reaction, the volume is found from the formula

$$\begin{aligned} V &= \frac{\dot{Q}}{k}\ln\frac{C_{\text{in}}}{C_{\text{out}}} \\ &= \left(\frac{660{,}430\ \frac{\text{gal}}{\text{day}}}{465\text{ day}^{-1}}\right)\left(\ln\frac{0.00104\ \frac{\text{lbm}}{\text{gal}}}{0.000015\ \frac{\text{lbm}}{\text{gal}}}\right) \\ &= 6020.4597\text{ gal} \quad (6.0\times 10^3\text{ gal}) \end{aligned}$$

The answer is (C).

50. The volume of each reactor is

$$V_1 = V_2 = V_3 = 500\text{ m}^3$$

The volumetric flow rates of species M and N are

$$\dot{V}_{\text{M},0} = \dot{V}_{\text{N},0} = 20\text{ m}^3/\text{min}$$

The reaction rate constant is

$$k = 0.078\text{ m}^3/\text{kmol·min}$$

The space-time is

$$\begin{aligned} \tau &= \frac{V}{\dot{V}_{\text{M},0} + \dot{V}_{\text{N},0}} \\ &= \frac{500\text{ m}^3}{20\ \frac{\text{m}^3}{\text{min}} + 20\ \frac{\text{m}^3}{\text{min}}} \\ &= 12.5\text{ min} \end{aligned}$$

The initial concentrations of species M and N into the first reactor are

$$C_{M,0} = C_{N,0} = 3.5\ \text{kmol/m}^3$$

The reaction is second-order liquid phase and is carried out in a CSTR. From *NCEES Handbook:* Second-Order Reactions, the space-time (in terms of the conversion in the first reactor, reaction rate constant, and initial concentration of species M) is

$$k\tau = \frac{X_A}{C_{A,0}(1 - X_A)^2}$$

Applying the preceding equation to the problem statement and solving for the space time gives

$$\tau = \frac{X_1}{kC_{M,0}(1 - X_1)^2}$$

Solving the space-time equation for the conversion gives

$$\begin{aligned}\frac{X_1}{(1 - X_1)^2} &= \tau k C_{M,0} \\ &= (12.5\ \text{min})\left(0.078\ \frac{\text{m}^3}{\text{kmol·min}}\right)\left(3.5\ \frac{\text{kmol}}{\text{m}^3}\right) \\ &= 3.4125 \\ X_1 &= (3.4125)(1 - X_1)^2 \\ &= 3.4125X_1^2 - 6.8250X_1 + 3.4125\end{aligned}$$

$$3.4125X_1^2 - 7.8250X_1 + 3.4125 = 0$$

From *NCEES Handbook:* Polynomials, the quadratic formula gives

$$\begin{aligned}X_1 &= \frac{-b \pm \sqrt{b^2 - 4ac}}{2a} \\ &= \frac{-(-7.8250) \pm \sqrt{(-7.8250)^2 - (4)(3.4125)(3.4125)}}{(2)(3.4125)} \\ &= \frac{7.8250 \pm 3.828}{6.8250} \\ &= 1.707 \text{ or } 0.586\end{aligned}$$

The fraction of species M converted in the first reactor, X_1, must be between zero and one, so it is 0.586.

From *NCEES Handbook:* Reaction Parameters, the fractional conversion is

$$X_A = \frac{C_{Ao} - C_A}{C_{Ao}}$$

Applying the preceding equation to the conditions of the problem statement and solving for the concentration of the species M from the first reactor gives

$$\begin{aligned}C_{M,1} &= C_{M,0}(1 - X_1) \\ &= \left(3.5\ \frac{\text{kmol}}{\text{m}^3}\right)(1 - 0.586) \\ &= 1.45\ \text{kmol/m}^3\end{aligned}$$

This is also the concentration of the species M into the second reactor. The reaction is second-order liquid phase and is carried out in a CSTR, so the space-time (in terms of the conversion in the second reactor, reaction rate constant, and concentration of species M in the second reactor) is

$$\tau = \frac{X_2}{kC_{M,1}(1 - X_2)^2}$$

Solving the space-time equation for the conversion in the second reactor gives

$$\begin{aligned}\frac{X_2}{(1 - X_2)^2} &= \tau k C_{M,1} \\ &= (12.5\ \text{min})\left(0.078\ \frac{\text{m}^3}{\text{kmol·min}}\right)\left(1.45\ \frac{\text{kmol}}{\text{m}^3}\right) \\ &= 1.4138 \\ X_2 &= (1.4138)(1 - X_2)^2 \\ &= 1.4138X_2^2 - 2.8275X_2 + 1.4138\end{aligned}$$

$$1.4138X_2^2 - 3.8275X_2 + 1.4138 = 0$$

The quadratic formula gives

$$\begin{aligned}X_2 &= \frac{-b \pm \sqrt{b^2 - 4ac}}{2a} \\ &= \frac{-(-3.8275) \pm \sqrt{(-3.8275)^2 - (4)(1.4138)(1.4138)}}{(2)(1.4138)} \\ &= \frac{3.8275 \pm 2.580}{2.8276} \\ &= 2.266 \text{ or } 0.441\end{aligned}$$

The conversion in the second reactor, X_2, must be between zero and one, so it is 0.441.

The concentration of species M from the second reactor is

$$\begin{aligned}C_{M,2} &= C_{M,1}(1 - X_2) \\ &= \left(1.45\ \frac{\text{kmol}}{\text{m}^3}\right)(1 - 0.441) \\ &= 0.811\ \text{kmol/m}^3\end{aligned}$$

This is also the concentration of species M into the third reactor. For the third reactor, the space-time is

$$\tau = \frac{X_3}{kC_{M,2}(1 - X_3)^2}$$

Solving the space-time equation for the conversion in the third reactor gives

$$\begin{aligned}
\frac{X_3}{(1 - X_3)^2} &= \tau k C_{M,2} \\
&= (12.5 \text{ min})\left(0.078 \ \frac{\text{m}^3}{\text{kmol·min}}\right)\left(0.811 \ \frac{\text{kmol}}{\text{m}^3}\right) \\
&= 0.7907 \\
X_3 &= (0.7907)(1 - X_3)^2 \\
&= 0.7907X_3^2 - 1.5815X_3 + 0.7907
\end{aligned}$$

$$0.7907X_3^2 - 2.5815X_3 + 0.7907 = 0$$

The quadratic formula gives

$$\begin{aligned}
X_3 &= \frac{-b \pm \sqrt{b^2 - 4ac}}{2a} \\
&= \frac{-(-2.5815) \pm \sqrt{(-2.5815)^2 - (4)(0.7907)(0.7907)}}{(2)(0.7907)} \\
&= \frac{2.5815 \pm 2.040}{1.5815} \\
&= 2.922 \text{ or } 0.342
\end{aligned}$$

The conversion in the third reactor, X_3, must be between zero and one, so it is 0.342.

The concentration of species M from the third reactor is

$$\begin{aligned}
C_{M,3} &= C_{M,2}(1 - X_3) \\
&= \left(0.811 \ \frac{\text{kmol}}{\text{m}^3}\right)(1 - 0.342) \\
&= 0.534 \text{ kmol/m}^3
\end{aligned}$$

The net conversion of species M in the series of the three CSTRs is related to the concentration of species M entering the first reactor and the concentration of species M leaving the third reactor by the expression

$$\begin{aligned}
X_{\text{net}} &= \frac{C_{M,0} - C_{M,3}}{C_{M,0}} \\
&= \frac{3.5 \ \frac{\text{kmol}}{\text{m}^3} - 0.534 \ \frac{\text{kmol}}{\text{m}^3}}{3.5 \ \frac{\text{kmol}}{\text{m}^3}} \\
&= 0.847 \quad (0.80)
\end{aligned}$$

The answer is (D).

51. Start by finding how much of each component is in the combustion products. This solution will base the calculations on 1 mol of fuel gas entering the combustion chamber. The fuel gas is 25% carbon monoxide (CO) and 75% nitrogen (N_2), so 1 mol of fuel gas contains

$$\begin{aligned}
n_{\text{CO,fuel}} &= (0.25)(1 \text{ mol}) \\
&= 0.25 \text{ mol} \\
n_{\text{N}_2\text{,fuel}} &= (0.75)(1 \text{ mol}) \\
&= 0.75 \text{ mol}
\end{aligned}$$

The combustion reaction is

$$2\text{CO} + \text{O}_2 \rightarrow 2\text{CO}_2$$

All the carbon in the CO in the fuel gas ends up in the carbon dioxide (CO_2) in the combustion products, and no carbon is contributed by the air, so the number of moles of CO_2 produced by the combustion, $n_{\text{CO}_2\text{,out}}$, is the same as the number of moles of CO in the fuel gas, $n_{\text{CO,fuel}}$. For 1 mol of fuel gas,

$$n_{\text{CO}_2\text{,out}} = n_{\text{CO,fuel}} = 0.25 \text{ mol}$$

The additional oxygen that reacts with the CO to create CO_2 is supplied by the air. From the stoichiometry of the reaction, the number of moles of O_2 needed to react completely with the CO present in the fuel gas is

$$n_{\text{O}_2\text{,needed}} = 0.5n_{\text{CO,fuel}} = (0.5)(0.25 \text{ mol}) = 0.125 \text{ mol}$$

The excess air is 160%, so the number of moles of oxygen in the excess air is

$$n_{\text{O}_2\text{,excess}} = 1.6n_{\text{O}_2\text{,needed}} = (1.6)(0.125 \text{ mol}) = 0.2 \text{ mol}$$

The number of moles of oxygen in the total inlet air, then, that is used with 1 mol of fuel gas is

$$\begin{aligned}
n_{\text{O}_2\text{,air}} &= n_{\text{O}_2\text{,needed}} + n_{\text{O}_2\text{,excess}} \\
&= 0.125 \text{ mol} + 0.2 \text{ mol} \\
&= 0.325 \text{ mol}
\end{aligned}$$

From *NCEES Handbook:* Material Balances, the mass balance for the excess oxygen is

$$\begin{aligned} \textit{Accumulation} &= \textit{Output} - \textit{Input} \\ &\quad + \textit{Generation} - \textit{Consumption} \end{aligned}$$

The excess O_2 leaves the combustion chamber unchanged, and there is no other source of O_2, so

$$n_{\mathrm{O_2,out}} = n_{\mathrm{O_2,excess}} = 0.2 \text{ mol}$$

The air consists of 21% (by mole) O_2, so the total inlet air is

$$n_{\mathrm{air}} = \frac{n_{\mathrm{O_2,air}}}{0.21} = \frac{0.325 \text{ mol}}{0.21} = 1.548 \text{ mol}$$

The other 79% (by mole) of the air is N_2, so

$$n_{\mathrm{N_2,air}} = 0.79 n_{\mathrm{air}} = (0.79)(1.548 \text{ mol}) = 1.223 \text{ mol}$$

N_2 enters the combustion chamber in both the air and the fuel gas and does not react, so it leaves the chamber unchanged. From *NCEES Handbook:* Material Balances, the mass balance for the nitrogen is

$$\begin{aligned} \textit{Accumulation} &= \textit{Output} - \textit{Input} \\ &\quad + \textit{Generation} - \textit{Consumption} \end{aligned}$$

A mass balance of the N_2 gives

$$\begin{aligned} n_{\mathrm{N_2,out}} &= n_{\mathrm{N_2,air}} + n_{\mathrm{N_2,fuel}} \\ &= 1.223 \text{ mol} + 0.75 \text{ mol} \\ &= 1.973 \text{ mol} \end{aligned}$$

For each 1 mol of fuel gas, then, the inlet air will be 1.548 mol, and the combustion products will be 0.25 mol of CO_2, 0.2 mol of O_2, and 1.973 mol of N_2. From *NCEES Handbook:* Heat of Reaction, the reaction enthalpy is

$$\Delta \hat{h}_R^0 = \sum_{\text{products}} \Delta \hat{h}_f^0 - \sum_{\text{reactants}} \Delta \hat{h}_f^0$$

At 25°C, the enthalpy change for the combustion reaction is

$$\begin{aligned} \Delta h_R &= \Delta h_{f,\mathrm{CO_2}} - \Delta h_{f,\mathrm{CO}} \\ &= -94.052 \ \frac{\text{kcal}}{\text{mol}} - \left(-26.412 \ \frac{\text{kcal}}{\text{mol}}\right) \\ &= -67.640 \text{ kcal/mol} \end{aligned}$$

The negative sign indicates that the reaction produces heat (rather than absorbing heat). The quantity of heat produced by the combustion of 1 mol of fuel gas is

$$\begin{aligned} Q &= n_{\mathrm{CO,fuel}}(-\Delta h_R) \\ &= (0.25 \text{ mol})\left(-\left(-67.640 \ \frac{\text{kcal}}{\text{mol}}\right)\right) \\ &= 16.910 \text{ kcal} \end{aligned}$$

Because the combustion chamber is adiabatic, this heat is consumed totally by the gases. The flame temperature is T_{flame}. From *NCEES Handbook:* Latent and Sensible Heat, the heat transferred is

$$\dot{Q}_{\mathrm{sensible}} = \dot{m} c_{\mathrm{p}} \Delta T$$

Applying the preceding equation to the conditions of the problem statement gives the energy balance around the combustion chamber.

$$\begin{aligned} Q &= n_{\mathrm{CO_2,out}} c_{p,\mathrm{CO_2}}(T_{\mathrm{flame}} - T_{\mathrm{in}}) \\ &\quad + n_{\mathrm{O_2,out}} c_{p,\mathrm{O_2}}(T_{\mathrm{flame}} - T_{\mathrm{in}}) \\ &\quad + n_{\mathrm{N_2,out}} c_{p,\mathrm{N_2}}(T_{\mathrm{flame}} - T_{\mathrm{in}}) \end{aligned}$$

c_p is the molar heat capacity. The fuel gas and air both enter the combustion chamber at $T_{\mathrm{in}} = 25°\mathrm{C}$, so solving for the flame temperature gives

$$\begin{aligned} T_{\mathrm{flame}} &= \frac{Q}{\left(\begin{array}{l} n_{\mathrm{CO_2,out}} c_{p,\mathrm{CO_2}} \\ \quad + n_{\mathrm{O_2,out}} c_{p,\mathrm{O_2}} \\ \quad + n_{\mathrm{N_2,out}} c_{p,\mathrm{N_2}} \end{array}\right)} + T_{\mathrm{in}} \\ &= \frac{(16.910 \text{ kcal})\left(1000 \ \frac{\text{cal}}{\text{kcal}}\right)}{\left(\begin{array}{l} (0.25 \text{ mol})\left(12.00 \ \frac{\text{cal}}{\text{mol·°C}}\right) \\ \quad + (0.2 \text{ mol})\left(7.90 \ \frac{\text{cal}}{\text{mol·°C}}\right) \\ \quad + (1.973 \text{ mol})\left(7.55 \ \frac{\text{cal}}{\text{mol·°C}}\right) \end{array}\right)} + 25°\mathrm{C} \\ &= 893.24°\mathrm{C} \quad (890°\mathrm{C}) \end{aligned}$$

The answer is (D).

52. The reaction is given as

$$\mathrm{PH_3} \rightarrow \mathrm{P_4} + \mathrm{H_2}$$

A decomposition reaction is usually first order. By inspection, this is a first-order reaction. In addition, the plot of $1/T$ versus $\ln k$ is a straight line, which further confirms that the reaction is first order.

$\ln k$	$1/T$ (K^{-1})
–5.3475076	0.0011765
–2.7806209	0.0010549
–1.3903024	0.0010000
2.29958058	0.0008333
5.70044357	0.0006667

Arrhenius' law relates any pair of reaction rate constants to their respective temperatures. This relationship allows the calculation of the activation energy, E, for the reaction from the rate constants for any two temperatures.

At a temperature of 948K, the reaction rate constant k_1 is given as 0.0620 s^{-1}. At a temperature of 1000K, the reaction rate constant k_2 is given as 0.249 s^{-1}.

From *NCEES Handbook:* Temperature Dependence, the activation energy is

$$\begin{aligned} E_a &= \frac{RT_1T_2}{(T_1 - T_2)}\ln\left(\frac{k_2}{k_1}\right) \\ &= \frac{\left(1.987\ \frac{\text{cal}}{\text{mol·K}}\right)(948\text{K})(1000\text{K})}{948\text{K} - 1000\text{K}}\ln\frac{0.0620\ \text{s}^{-1}}{0.249\ \text{s}^{-1}} \\ &= 50\,364\ \text{cal/mol} \quad (50\,000\ \text{cal/mol}) \end{aligned}$$

The answer is (B).

53. From *NCEES Handbook:* Material Balances, the balance equation is

$$\begin{aligned} \textit{Accumulation} &= \textit{Input} - \textit{Output} \\ &\quad + \textit{Generation} - \textit{Consumption} \end{aligned}$$

A total mass balance around the system gives

$$F = V + L$$

From *NCEES Handbook:* Distribution of Components Between Phases in a Vapor/Liquid Equilibrium, the distribution coefficient is defined as

$$K_i = \frac{y_i}{x_i}$$

For benzene, the ratio of the vapor mole fraction, y, to the liquid mole fraction, x, is given as

$$\begin{aligned} K_{\text{benzene}} &= \frac{y_{\text{benzene}}}{x_{\text{benzene}}} \\ &= 1.748 \end{aligned}$$

For toluene, the same ratio is given as

$$\begin{aligned} K_{\text{toluene}} &= \frac{y_{\text{toluene}}}{x_{\text{toluene}}} \\ &= 0.6223 \end{aligned}$$

Benzene is 47% of the feed, F, so a mass balance of benzene around the system gives

$$\begin{aligned} 0.47F &= y_{\text{benzene}}V + x_{\text{benzene}}L \\ &= (K_{\text{benzene}}x_{\text{benzene}})V + x_{\text{benzene}}L \\ &= x_{\text{benzene}}(K_{\text{benzene}}V + L) \end{aligned}$$

The liquid mole fraction of benzene is

$$\begin{aligned} x_{\text{benzene}} &= \frac{0.47F}{K_{\text{benzene}}V + L} \\ &= \frac{0.47F}{K_{\text{benzene}}V + (F - V)} \\ &= \frac{0.47F}{F + V(K_{\text{benzene}} - 1)} \end{aligned}$$

Similarly, toluene is 53% of the feed, so the liquid mole fraction of toluene is

$$x_{\text{toluene}} = \frac{0.53F}{F + V(K_{\text{toluene}} - 1)}$$

The mixture contains no other components, so the total mass of the liquid is

$$\begin{aligned} x_{\text{benzene}} + x_{\text{toluene}} &= \frac{0.47F}{F + V(K_{\text{benzene}} - 1)} \\ &\quad + \frac{0.53F}{F + V(K_{\text{toluene}} - 1)} \\ &= 1.0 \end{aligned}$$

Following similar steps, find the vapor mole fraction of the benzene.

$$\begin{aligned} 0.47F &= y_{\text{benzene}}V + x_{\text{benzene}}L \\ &= y_{\text{benzene}}V + \left(\frac{y_{\text{benzene}}}{K_{\text{benzene}}}\right)(F - V) \\ 0.47FK_{\text{benzene}} &= y_{\text{benzene}}VK_{\text{benzene}} + y_{\text{benzene}}(F - V) \\ y_{\text{benzene}} &= \frac{0.47FK_{\text{benzene}}}{F + V(K_{\text{benzene}} - 1)} \end{aligned}$$

Similarly,

$$y_{\text{toluene}} = \frac{0.53FK_{\text{toluene}}}{F + V(K_{\text{toluene}} - 1)}$$

The total mass of the vapor is

$$y_{benzene} + y_{toluene} = \frac{0.47FK_{benzene}}{F + V(K_{benzene} - 1)} + \frac{0.53FK_{toluene}}{F + V(K_{toluene} - 1)} = 1.0$$

As the total masses of the liquid and vapor are both equal to one, subtracting one from the other gives zero.

$$\left(\frac{0.47F}{F + V(K_{benzene} - 1)} + \frac{0.53F}{F + V(K_{toluene} - 1)}\right) - \left(\frac{0.47FK_{benzene}}{F + V(K_{benzene} - 1)} + \frac{0.53FK_{toluene}}{F + V(K_{toluene} - 1)}\right) = 0$$

Simplifying,

$$\frac{0.47(1 - K_{benzene})}{F + V(K_{benzene} - 1)} + \frac{0.53(1 - K_{toluene})}{F + V(K_{toluene} - 1)} = 0$$

$$0.47(1 - K_{benzene})\big(F + V(K_{toluene} - 1)\big) + 0.53(1 - K_{toluene})\big(F + V(K_{benzene} - 1)\big) = 0$$

Solving for the molar vapor flow rate gives

$$V = \frac{F(1 - 0.47K_{benzene} - 0.53K_{toluene})}{(K_{benzene} - 1)(K_{toluene} - 1)} = \frac{\left(1\ \frac{\text{mol}}{\text{h}}\right)(1 - (0.47)(1.748) - (0.53)(0.6223))}{(1.748 - 1)(0.6223 - 1)} = 0.5358\ \text{kmol/h}$$

The ratio of the molar vapor flow rate to the molar liquid flow rate is

$$\text{ratio} = \frac{V}{L} = \frac{V}{F - V} = \frac{0.5358\ \frac{\text{mol}}{\text{h}}}{1\ \frac{\text{mol}}{\text{h}} - 0.5358\ \frac{\text{mol}}{\text{h}}} = 1.154\quad (1.2)$$

The answer is (C).

54. The relevant data are shown in the illustration.

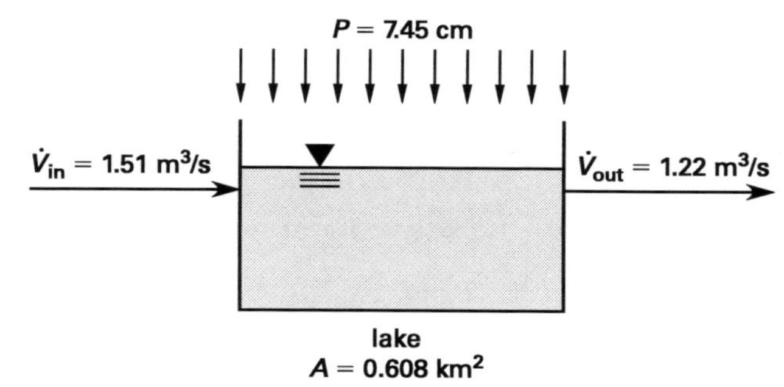

From *NCEES Handbook:* Material Balances, the material balance is

$$Accumulation = Input - Output + Generation - Consumption$$

The input to the lake consists of the inflow and the precipitation. The output consists of the outflow and the evaporation. The resulting change in volume (accumulation), ΔV, is 750 000 m^3. A water mass balance around the lake gives

$$\Delta V = V_{in} + V_{precip} - V_{out} - V_{evap}$$

The time, t, is 30 d. The volume of inflow is

$$V_{in} = \dot{V}_{in}t = \left(1.51\ \frac{\text{m}^3}{\text{s}}\right)(30\ \text{d})\left(86\,400\ \frac{\text{s}}{\text{d}}\right) = 3\,913\,920\ \text{m}^3$$

The volume of water the lake received from precipitation is

$$V_{precip} = PA = \left(\frac{7.45\ \text{cm}}{100\ \frac{\text{cm}}{\text{m}}}\right)(0.608\ \text{km}^2)\left(1000\ \frac{\text{m}}{\text{km}}\right)^2 = 45\,296\ \text{m}^3$$

The volume of outflow is

$$V_{out} = \dot{V}_{out}t = \left(1.22\ \frac{\text{m}^3}{\text{s}}\right)(30\ \text{d})\left(86\,400\ \frac{\text{s}}{\text{d}}\right) = 3\,162\,240\ \text{m}^3$$

So,

$$\begin{aligned}\Delta V &= V_{\text{in}} + V_{\text{precip}} - V_{\text{out}} - V_{\text{evap}} \\ V_{\text{evap}} &= V_{\text{in}} + V_{\text{precip}} - V_{\text{out}} - \Delta V \\ &= 3\,913\,920 \text{ m}^3 + 45\,296 \text{ m}^3 \\ &\quad -3\,162\,240 \text{ m}^3 - 750\,000 \text{ m}^3 \\ &= 46\,976 \text{ m}^3\end{aligned}$$

Divide the volume of evaporation by the surface area to find the evaporation depth.

$$\begin{aligned}d &= \frac{V_{\text{evap}}}{A} \\ &= \frac{(46\,976 \text{ m}^3)\left(100 \frac{\text{cm}}{\text{m}}\right)^3}{(0.608 \text{ km}^2)\left(100\,000 \frac{\text{cm}}{\text{km}}\right)^2} \\ &= 7.726 \text{ cm} \quad (7.7 \text{ cm})\end{aligned}$$

The answer is (B).

55. The diameter of the tube is given as

$$\begin{aligned}D &= \frac{10 \text{ mm}}{1000 \frac{\text{mm}}{\text{m}}} \\ &= 0.010 \text{ m}\end{aligned}$$

The air mass flow rate is given as

$$\begin{aligned}\dot{m} &= \frac{0.015 \frac{\text{kg}}{\text{min}}}{60 \frac{\text{s}}{\text{min}}} \\ &= 0.000\,25 \text{ kg/s}\end{aligned}$$

The air's absolute viscosity, μ, is 1.846×10^{-5} N·s/m². The air's kinematic viscosity, ν, is 1.67×10^{-5} m²/s. The diffusivity of the ethyl acetate in air, D_{AB}, is 8.9×10^{-6} m²/s. From *NCEES Handbook* table "Dimensionless Numbers," the Reynolds number is

$$\begin{aligned}\text{Re} &= \frac{\rho u D}{\mu} = \frac{\rho\left(\frac{\dot{Q}}{A}\right)D}{\mu} = \frac{\rho \dot{Q} D}{\left(\frac{\pi D^2}{4}\right)\mu} \\ &= \frac{4\dot{m}}{\pi D \mu} \\ &= \frac{(4)\left(0.000\,25 \frac{\text{kg}}{\text{s}}\right)}{\pi(0.010 \text{ m})\left(1.846 \times 10^{-5} \frac{\text{N·s}}{\text{m}^2}\right)} \\ &= 1724.32\end{aligned}$$

Because the Reynolds number is less than 2000, the flow of air can be considered laminar. From *NCEES Handbook* table "Dimensionless Numbers," the Schmidt number is

$$\begin{aligned}\text{Sc} &= \frac{\nu}{D_{AB}} = \frac{1.67 \times 10^{-5} \frac{\text{m}^2}{\text{s}}}{8.9 \times 10^{-6} \frac{\text{m}^2}{\text{s}}} \\ &= 1.876\end{aligned}$$

To decide what Sherwood number to use, evaluate the following expression.

$$\left(\frac{(\text{Re})(\text{Sc})}{\frac{L}{D}}\right)^{1/3} = \left(\frac{(1724.32)(1.876)}{\frac{1 \text{ m}}{0.010 \text{ m}}}\right)^{1/3} = 3.186$$

Because the preceding value is larger than 2, the Sherwood number to use is

$$\begin{aligned}\text{Sh} &= 1.86\left(\frac{(\text{Re})(\text{Sc})}{\frac{L}{D}}\right)^{1/3} \\ &= (1.86)(3.186) \\ &= 5.926\end{aligned}$$

From *NCEES Handbook* table "Dimensionless Numbers," the Sherwood number is

$$\text{Sh} = \frac{h_m L}{D_{AB}}$$

Solving the preceding equation for the average mass transfer convection coefficient and applying it to the conditions of the problem statement gives

$$h_m = (\text{Sh})\left(\frac{D_{AB}}{D}\right) = (5.926)\left(\frac{8.9 \times 10^{-6}\ \frac{\text{m}^2}{\text{s}}}{0.010\ \text{m}}\right) = 0.005\,27\ \text{m/s} \quad (5.3 \times 10^{-3}\ \text{m/s})$$

The answer is (B).

56. The diameter of the pipe is

$$D = \frac{2.469\ \text{in}}{12\ \frac{\text{in}}{\text{ft}}} = 0.2058\ \text{ft}$$

The cross-sectional area of the pipe is

$$A = \frac{\pi D^2}{4} = \frac{\pi(0.2058\ \text{ft})^2}{4} = 0.03326\ \text{ft}^2$$

The velocity of the water through the pipe is

$$u = \frac{\dot{Q}}{A} = \frac{\dfrac{62\ \frac{\text{gal}}{\text{min}}}{\left(7.48\ \frac{\text{gal}}{\text{ft}^3}\right)\left(60\ \frac{\text{sec}}{\text{min}}\right)}}{0.03326\ \text{ft}^2} = 4.15\ \text{ft/sec}$$

The density of the water is given as

$$\rho = 62.2\ \text{lbm/ft}^3$$

The viscosity of the water is

$$\mu = (0.86\ \text{cP})\left(6.72 \times 10^{-4}\ \frac{\frac{\text{lbm}}{\text{ft-sec}}}{\text{cP}}\right) = 0.0005779\ \text{lbm/ft-sec}$$

From *NCEES Handbook* table "Dimensionless Numbers," the Reynolds number is

$$Re = \frac{\rho u D}{\mu} = \frac{\left(62.2\ \frac{\text{lbm}}{\text{ft}^3}\right)\left(4.15\ \frac{\text{ft}}{\text{sec}}\right)(0.2058\ \text{ft})}{0.0005779\ \frac{\text{lbm}}{\text{ft-sec}}} = 91{,}900$$

The roughness of the pipe is given as

$$\varepsilon = 0.0002\ \text{ft}$$

From *NCEES Handbook:* Absolute Roughness and Relative Roughness, the relative roughness of the pipe is

$$\frac{\varepsilon}{D} = \frac{0.0002\ \text{ft}}{0.2058\ \text{ft}} = 0.0009718$$

With the Reynolds number and the relative roughness, the friction factor can be determined from the Moody chart.

$$f = 0.0223$$

The equivalent length of the pipe, L, is 225 ft. From *NCEES Handbook:* Head Loss in Pipe or Conduit, the Darcy-Weisbach equation gives the head loss due to friction between points 1 and 2 is

$$h_L = f\frac{L}{D}\frac{u^2}{2g} = (0.0223)\frac{(225\ \text{ft})\left(4.15\ \frac{\text{ft}}{\text{sec}}\right)^2}{(0.2058\ \text{ft})(2)\left(32.2\ \frac{\text{ft}}{\text{sec}^2}\right)} = 6.52\ \text{ft}$$

As in *NCEES Handbook:* The Bernoulli Equation, the Bernoulli equation for flow from point 1 to point 2, with friction and no pump work, is

$$\frac{P_1 g_c}{\rho g} + \frac{u_1^2}{2g} + z_1 = \frac{P_2 g_c}{\rho g} + \frac{u_2^2}{2g} + z_2 + h_L$$

The velocity of the fluid is the same at points 1 and 2, so

$$\frac{P_1 g_c}{\rho g} + z_1 = \frac{P_2 g_c}{\rho g} + z_2 + h_L$$

From the illustration, the static head between point 1 and 2 is

$$z_2 - z_1 = 25\ \text{ft}$$

Solving for the pressure at point 1,

$$\begin{aligned} P_1 &= P_2 + (z_2 - z_1 + h_L)\left(\frac{\rho g}{g_c}\right) \\ &= \left(18\ \frac{\text{lbf}}{\text{in}^2}\right)\left(12\ \frac{\text{in}}{\text{ft}}\right)^2 + (25\ \text{ft} + 6.52\ \text{ft}) \\ &\quad \times \left(\frac{\left(62.2\ \frac{\text{lbm}}{\text{ft}^3}\right)\left(32.2\ \frac{\text{ft}}{\text{sec}^2}\right)}{32.2\ \frac{\text{ft-lbm}}{\text{lbf-sec}^2}}\right) \\ &= 4553\ \text{lbf/ft}^2 \end{aligned}$$

The level of the fluid in the surge tank relative to the ground level is

$$\begin{aligned} h &= \frac{P_1 g_c}{\rho g} \\ &= \frac{\left(4553\ \frac{\text{lbf}}{\text{ft}^2}\right)\left(32.2\ \frac{\text{ft-lbm}}{\text{lbf-sec}^2}\right)}{\left(62.2\ \frac{\text{lbm}}{\text{ft}^3}\right)\left(32.2\ \frac{\text{ft}}{\text{sec}^2}\right)} \\ &= 73.20\ \text{ft} \end{aligned}$$

The base of the surge tank is 25 ft above ground level, so the depth of the water in the tank is

$$\begin{aligned} d &= h - 25\ \text{ft} \\ &= 73.20\ \text{ft} - 25\ \text{ft} \\ &= 48.20\ \text{ft} \quad (48\ \text{ft}) \end{aligned}$$

The answer is (B).

57. The mass flow rate of activated sludge treated, $\dot{m}_F$, is 2300 lbm/day. The absolute temperature of the air entering the system is

$$T = 72°\text{F} + 460° = 532°\text{R}$$

The pressure of the air entering the system, P, is 14.7 lbf/in^2. The standard temperature is

$$T_{\text{ideal}} = 32°\text{F} + 460° = 492°\text{R}$$

From *NCEES Handbook* table "Standard Values," the molar standard volume of the ideal gas at standard conditions (STP) is 359 ft^3/lb mole. From *NCEES Handbook* table "Selected Properties of Air," the molecular mass of air, MW, is 28.965 lbm/lb mole. Because the sludge contains 82% volatiles, the mass of volatile solids in the incoming sludge is

$$\begin{aligned} \dot{m}_s &= 0.82\dot{m}_F = (0.82)\left(2300\ \frac{\text{lbm}}{\text{day}}\right) \\ &= 1886\ \text{lbm/day} \end{aligned}$$

Because 63% of solids in the waste-activated sludge are destroyed, the mass of volatile solids destroyed is

$$\begin{aligned} \dot{m}_d &= 0.63\dot{m}_s \\ &= (0.63)\left(1886\ \frac{\text{lbm}}{\text{day}}\right) \\ &= 1188.18\ \text{lbm/day} \end{aligned}$$

2.0 lbm of oxygen is needed per 1 lbm of solid, so the mass of oxygen needed is

$$\begin{aligned} \dot{m}_{O_2} &= \left(\frac{2.0\ \text{lbm}}{1\ \text{lbm}}\right)\dot{m}_d \\ &= \left(\frac{2.0\ \text{lbm}}{1\ \text{lbm}}\right)\left(1188.18\ \frac{\text{lbm}}{\text{day}}\right) \\ &= 2376.36\ \text{lbm/day} \end{aligned}$$

From *NCEES Handbook:* Density and Relative Density, the density is

$$\rho = \frac{m}{V}$$

From *NCEES Handbook:* Ideal Gas Law, the ideal gas law is

$$PV = \frac{mRT}{MW}$$

Using the ideal gas law, the density of the air used is

$$p = \frac{m}{V} = MW\frac{n}{V}$$
$$= MW\left(\frac{n_{\text{ideal}}}{V_{\text{ideal}}}\right)\left(\frac{P}{P_{\text{ideal}}}\right)\left(\frac{T_{\text{ideal}}}{T}\right)$$
$$= MW\left(\frac{1}{V_{\text{ideal}}}\right)\left(\frac{14.7\ \frac{\text{lbf}}{\text{in}^2}}{14.7\ \frac{\text{lbf}}{\text{in}^2}}\right)\left(\frac{T_{\text{ideal}}}{T}\right)$$
$$= \frac{MWT_{\text{ideal}}}{V_{\text{ideal}}T}$$
$$= \frac{\left(28.965\ \frac{\text{lbm}}{\text{lb mole}}\right)(492^\circ\text{R})}{\left(359\ \frac{\text{ft}^3}{\text{lb mole}}\right)(532^\circ\text{R})}$$
$$= 0.07462\ \text{lbm/ft}^3$$

Because the air contains 23.3% oxygen, the volume of air needed is

$$V = \frac{\dot{m}_{O_2}}{0.233\rho}$$
$$= \frac{2376.36\ \frac{\text{lbm}}{\text{day}}}{(0.233)\left(0.07462\ \frac{\text{lbm}}{\text{ft}^3}\right)}$$
$$= 136{,}678.8\ \text{ft}^3/\text{day} \quad (140{,}000\ \text{ft}^3/\text{day})$$

The answer is (C).

58. Calculate the velocity at points 2, 3, and 4. The inside diameter of the pipe is

$$D = \frac{6.065\ \text{in}}{12\ \frac{\text{in}}{\text{ft}}}$$
$$= 0.5054\ \text{ft}$$

The cross-sectional area of the pipe is

$$A = \frac{\pi D^2}{4}$$
$$= \frac{\pi(0.5054\ \text{ft})^2}{4}$$
$$= 0.201\ \text{ft}^2$$

The velocity of the water at point 2 is

$$u_2 = \frac{\dot{Q}_2}{A}$$
$$= \frac{350\ \frac{\text{gal}}{\text{min}}}{(0.201\ \text{ft}^2)\left(7.48\ \frac{\text{gal}}{\text{ft}^3}\right)\left(60\ \frac{\text{sec}}{\text{min}}\right)}$$
$$= 3.88\ \text{ft/sec}$$

The velocity of the water at point 3 is

$$u_3 = \frac{\dot{Q}_3}{A} = \frac{\dot{Q}_2 + \dot{Q}_4}{A}$$
$$= \frac{350\ \frac{\text{gal}}{\text{min}} + 1590\ \frac{\text{gal}}{\text{min}}}{(0.201\ \text{ft}^2)\left(7.48\ \frac{\text{gal}}{\text{ft}^3}\right)\left(60\ \frac{\text{sec}}{\text{min}}\right)}$$
$$= 21.506\ \text{ft/sec}$$

The velocity of the water at point 4 is

$$u_4 = \frac{\dot{Q}_4}{A}$$
$$= \frac{1590\ \frac{\text{gal}}{\text{min}}}{(0.201\ \text{ft}^2)\left(7.48\ \frac{\text{gal}}{\text{ft}^3}\right)\left(60\ \frac{\text{sec}}{\text{min}}\right)}$$
$$= 17.63\ \text{ft/sec}$$

Calculate the Reynolds number and friction factor at points 2, 3, and 4. The density of the water, ρ, is 62.3 lbm/ft^3. The viscosity of the water is

$$\mu = (1.13\ \text{cP})\left(6.72 \times 10^{-4}\ \frac{\frac{\text{lbm}}{\text{ft-sec}}}{\text{cP}}\right)$$
$$= 0.000759\ \text{lbm/ft-sec}$$

From *NCEES Handbook* table "Dimensionless Numbers," the Reynolds number at point 2 is

$$Re_2 = \frac{\rho u_2 D}{\mu}$$
$$= \frac{\left(62.3\ \frac{\text{lbm}}{\text{ft}^3}\right)\left(3.88\ \frac{\text{ft}}{\text{sec}}\right)(0.5054\ \text{ft})}{0.000759\ \frac{\text{lbm}}{\text{ft-sec}}}$$
$$= 160{,}958$$

The Darcy friction factor at point 2 is

$$\begin{aligned} f_2 &= 0.00357 + 0.0218(Re_2)^{-0.55} \\ &= 0.00357 + (0.0218)(160{,}958)^{-0.55} \\ &= 0.00360 \end{aligned}$$

The Reynolds number at point 3 is

$$\begin{aligned} Re_3 &= \frac{\rho u_3 D}{\mu} \\ &= \frac{\left(62.3\ \frac{\text{lbm}}{\text{ft}^3}\right)\left(21.506\ \frac{\text{ft}}{\text{sec}}\right)(0.5054\ \text{ft})}{0.000759\ \frac{\text{lbm}}{\text{ft-sec}}} \\ &= 892{,}157 \end{aligned}$$

The Darcy friction factor at point 3 is

$$\begin{aligned} f_{L3} &= 0.00357 + 0.0218(Re_3)^{-0.55} \\ &= 0.00357 + (0.0218)(892{,}157)^{-0.55} \\ &= 0.00358 \end{aligned}$$

The Reynolds number at point 4 is

$$\begin{aligned} Re_4 &= \frac{\rho u_4 D}{\mu} \\ &= \frac{\left(62.3\ \frac{\text{lbm}}{\text{ft}^3}\right)\left(17.63\ \frac{\text{ft}}{\text{sec}}\right)(0.5054\ \text{ft})}{0.000759\ \frac{\text{lbm}}{\text{ft-sec}}} \\ &= 731{,}364 \end{aligned}$$

The Darcy friction factor at point 4 is

$$\begin{aligned} f_{L4} &= 0.00357 + 0.0218(Re_4)^{-0.55} \\ &= 0.00357 + (0.0218)(731{,}364)^{-0.55} \\ &= 0.00358 \end{aligned}$$

The total equivalent length of pipe from point 3 to point 2, L_2, is 650 ft. The water exits at point 2 and the K-value for an exit is 1.00. As in *NCEES Handbook:* Head Loss in Pipe or Conduit, the head loss due to friction at point 2 is

$$\begin{aligned} h_{L2} &= \left(\sum K + f_2 \frac{L_2}{D}\right)\left(\frac{u_2^2}{2g}\right) \\ &= \left(1.00 + (0.00360)\left(\frac{650\ \text{ft}}{0.5054\ \text{ft}}\right)\right)\left(\frac{\left(3.88\ \frac{\text{ft}}{\text{sec}}\right)^2}{(2)\left(32.2\ \frac{\text{ft}}{\text{sec}^2}\right)}\right) \\ &= 1.316\ \text{ft} \end{aligned}$$

The total equivalent length of pipe from the exit of the tank to point 3, L_3, is 2000 ft. The head loss due to friction at point 3 is

$$\begin{aligned} h_{L3} &= f_2 \frac{L_3 u_3^2}{D2g} \\ &= (0.00358)\left(\frac{(2000\ \text{ft})\left(21.506\ \frac{\text{ft}}{\text{sec}}\right)^2}{(0.5054\ \text{ft})(2)\left(32.2\ \frac{\text{ft}}{\text{sec}^2}\right)}\right) \\ &= 101.7\ \text{ft} \end{aligned}$$

The total equivalent length of pipe from point 3 to point 4, L_4, is 650 ft. The water exits at point 4, and the K-value for an exit is 1.00, so the head loss due to friction at point 4 is

$$\begin{aligned} h_{L4} &= \left(\sum K + f_4 \frac{L_4}{D}\right)\left(\frac{u_4^2}{2g}\right) \\ &= \left(1.00 + (0.00358)\left(\frac{650\ \text{ft}}{0.5054\ \text{ft}}\right)\right)\left(\frac{\left(17.63\ \frac{\text{ft}}{\text{sec}}\right)^2}{(2)\left(32.2\ \frac{\text{ft}}{\text{sec}^2}\right)}\right) \\ &= 27.05\ \text{ft} \end{aligned}$$

The total head loss due friction is

$$\begin{aligned} h_L &= h_{L2} + h_{L3} + h_{L4} \\ &= 1.316\ \text{ft} + 101.7\ \text{ft} + 27.05\ \text{ft} \\ &= 130.1\ \text{ft} \quad (130\ \text{ft}) \end{aligned}$$

The answer is (D).

59. The absolute temperature of the feed to the heater is

$$T_{\text{in}} = 30°\text{C} + 273° = 303\text{K}$$

The absolute temperature of the outlet stream is

$$T_{\text{out}} = 86°\text{C} + 273° = 359\text{K}$$

From *NCEES Handbook:* Heat Capacity/Specific Heat (c_p), the heat transferred is

$$\dot{Q} = \dot{m}c_p\Delta T$$

As the C_2H_4O in the mixture is heated from 303K to 359K, the change in its enthalpy is

$$\begin{aligned}
\Delta H_{C_2H_4O} &= \dot{m}_{C_2H_4O}\int_{303\text{K}}^{359\text{K}} c_{p,C_2H_4O}\,dT \\
&= \left(1\ \frac{\text{mol}}{\text{h}}\right)\int_{303\text{K}}^{359\text{K}} \left(\begin{array}{l} -3.2032\times10^{-3}\ \dfrac{\text{kJ}}{\text{mol·K}} \\ +\left(1.9521\times10^{-4}\ \dfrac{\text{kJ}}{\text{mol·K}^2}\right)T \\ -\left(7.7343\times10^{-8}\ \dfrac{\text{kJ}}{\text{mol·K}^3}\right)T^2 \end{array}\right) dT \\
&= \left(1\ \frac{\text{mol}}{\text{h}}\right)\left(\begin{array}{l} \left(-3.2032\times10^{-3}\ \dfrac{\text{kJ}}{\text{mol·K}}\right)T \\ +\dfrac{\left(1.9521\times10^{-4}\ \dfrac{\text{kJ}}{\text{mol·K}^2}\right)T^2}{2} \\ -\dfrac{\left(7.7343\times10^{-8}\ \dfrac{\text{kJ}}{\text{mol·K}^3}\right)T^3}{3} \end{array}\right)\Bigg|_{303\text{K}}^{359\text{K}} \\
&= \left(1\ \frac{\text{mol}}{\text{h}}\right)\left(\begin{array}{l} \left(-3.2032\times10^{-3}\ \dfrac{\text{kJ}}{\text{mol·K}}\right) \\ \quad\times(359\text{K}-303\text{K}) \\ +\dfrac{\left(\begin{array}{l}1.9521\times10^{-4}\ \dfrac{\text{kJ}}{\text{mol·K}^2} \\ \times\left((359\text{K})^2-(303\text{K})^2\right)\end{array}\right)}{2} \\ -\dfrac{\left(\begin{array}{l}7.7343\times10^{-8}\ \dfrac{\text{kJ}}{\text{mol·K}^3} \\ \times\left((359\text{K})^3-(303\text{K})^3\right)\end{array}\right)}{3} \end{array}\right) \\
&= 2.963\ \text{kJ/h}
\end{aligned}$$

As the H_2O in the mixture is heated from 303K to 359K, the change in its enthalpy is

$$\begin{aligned}
\Delta H_{H_2O} &= \dot{m}_{H_2O}\int_{303\text{K}}^{359\text{K}} c_{p,H_2O}\,dT \\
&= \left(4.8\ \frac{\text{mol}}{\text{h}}\right)\int_{303\text{K}}^{359\text{K}} \left(\begin{array}{l} 72.84\times10^{-3}\ \dfrac{\text{kJ}}{\text{mol·K}} \\ +\left(1.0400\times10^{-5}\ \dfrac{\text{kJ}}{\text{mol·K}^2}\right)T \\ -\left(1.4976\times10^{-9}\ \dfrac{\text{kJ}}{\text{mol·K}^3}\right)T^2 \end{array}\right) dT \\
&= \left(4.8\ \frac{\text{mol}}{\text{h}}\right)\left(\begin{array}{l} \left(72.84\times10^{-3}\ \dfrac{\text{kJ}}{\text{mol·K}}\right)T \\ +\dfrac{\left(1.0400\times10^{-5}\ \dfrac{\text{kJ}}{\text{mol·K}^2}\right)T^2}{2} \\ -\dfrac{\left(1.4976\times10^{-9}\ \dfrac{\text{kJ}}{\text{mol·K}^3}\right)T^3}{3} \end{array}\right)\Bigg|_{303\text{K}}^{359\text{K}} \\
&= \left(4.8\ \frac{\text{mol}}{\text{h}}\right)\left(\begin{array}{l} \left(72.84\times10^{-3}\ \dfrac{\text{kJ}}{\text{mol·K}}\right) \\ \quad\times(359\text{K}-303\text{K}) \\ +\dfrac{\left(\begin{array}{l}1.0400\times10^{-5}\ \dfrac{\text{kJ}}{\text{mol·K}^2} \\ \times\left((359\text{K})^2-(303\text{K})^2\right)\end{array}\right)}{2} \\ -\dfrac{\left(\begin{array}{l}1.4976\times10^{-9}\ \dfrac{\text{kJ}}{\text{mol·K}^3} \\ \times\left((359\text{K})^3-(303\text{K})^3\right)\end{array}\right)}{3} \end{array}\right) \\
&= 20.46\ \text{kJ/h}
\end{aligned}$$

The enthalpy change of the entire mixture as its temperature increases from 303K to 359K is

$$\begin{aligned}
\Delta H &= \Delta H_{C_2H_4O} + \Delta H_{H_2O} \\
&= 2.963\ \frac{\text{kJ}}{\text{h}} + 20.46\ \frac{\text{kJ}}{\text{h}} \\
&= 23.423\ \text{kJ/h}\quad (23\ \text{kJ/h})
\end{aligned}$$

The answer is (D).

60. The symbols matched to the correct labels are shown.

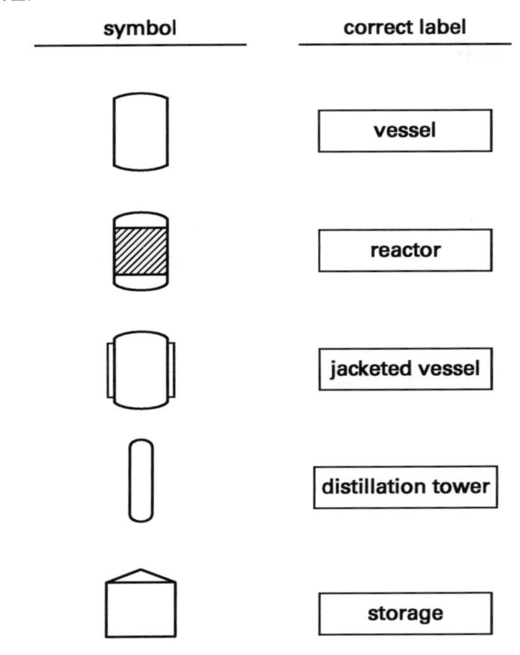

The correct answer choices are shown.

61. The reaction rate constant, k_1, is 0.002 min^{-1}. The reaction temperature is given as

$$T_1 = 260°\text{F} + 460° = 720°\text{R}$$

The activation energy, E_a, is 37,188.31 Btu/lb mole. The universal gas constant, R, has values of 1.987 Btu/lb mole-°R and 1545 ft-lbf/lb mole-°R. The reaction temperature is

$$T_2 = 140°\text{F} + 460° = 600°\text{R}$$

From *NCEES Handbook:* Temperature Dependence, the reaction rate constant at the reaction temperature is found from the relationship

$$\ln\left(\frac{k_1}{k_2}\right) = \frac{E_a}{R}\,\frac{T_1 - T_2}{T_1 T_2}$$

Solving for the reaction rate constant gives

$$\begin{aligned} k_2 &= k_1 e^{(E_a/R)((T_2 - T_1)/T_2T_1)} \\ &= 0.002\ \text{min}^{-1} e^{\left(\left(37{,}188.31\ \frac{\text{Btu}}{\text{lb mole}}\Big/1.986\ \frac{\text{Btu}}{\text{lb mole-°R}}\right)\times\left((720°\text{R}-600°\text{R})/(720°\text{R})(600°\text{R})\right)\right)} \\ &= 0.363\ \text{min}^{-1} \end{aligned}$$

The feed rate, F, is 25 lb mole/min. Because the feed consists of pure M, the mole fraction of species M in the feed, y, is 1. From *NCEES Handbook* table "Physical Constants," the gas constant is 1545 ft-lbf/lb mole-°R. From *NCEES Handbook:* Concentrations, the concentration is

$$c_i = x_i \frac{P}{RT}$$

Applying the preceding equation to the conditions of the problem statement the concentration of species M in the feed is

$$\begin{aligned} C_\text{M} &= \frac{yP}{RT_1} \\ &= \frac{(1)\left(118\ \frac{\text{lbf}}{\text{in}^2}\right)\left(12\ \frac{\text{in}}{\text{ft}}\right)^2}{\left(1545\ \frac{\text{ft-lbf}}{\text{lb mole-°R}}\right)(720°\text{R})} \\ &= 0.0153\ \text{lb mole/ft}^3 \end{aligned}$$

The desired conversion, X, is 0.85. From the stoichiometry of the reaction, 1 mol of M produces 1 mol of N and 2 mol of P. The stoichiometric coefficient difference for the reaction is the sum of the stoichiometric coefficients of the individual products (N and P) and the reactant (M). The coefficients of products are taken as positive and those of reactants are taken as negative.

$$\delta = 1 + 2 + (-1) = 2$$

From the stoichiometry of the reaction, the increase in the number of moles as the reaction progresses is

$$\varepsilon = y\delta = (1)(2) = 2$$

From *NCEES Handbook:* Rate Equations in Differential Form for Irreversible Reactions: First-Order, because the reaction is elementary, the reaction rate law is first order.

$$-r_\text{M} = k_2 C_\text{M}$$

From *NCEES Handbook:* First-Order Reactions, the following relationship applies.

$$\begin{aligned} k\tau &= -(1+\varepsilon_\text{A})\ln(1 - X_\text{A}) - \varepsilon_\text{A} X_\text{A} \\ \tau &= \frac{-(1+\varepsilon_\text{A})\ln(1 - X_\text{A}) - \varepsilon_\text{A} X_\text{A}}{k} \\ &= \frac{1}{k}\left((1+\varepsilon_\text{A})\ln\left(\frac{1}{1 - X_\text{A}}\right) - \varepsilon_\text{A} X_\text{A}\right) \end{aligned}$$

From *NCEES Handbook:* Plug-Flow Reactor,

$$\tau = \frac{C_{\text{Ao}} V_{\text{PFR}}}{F_{\text{Ao}}}$$

Replacing,

$$\tau = \frac{C_{\text{Ao}}V_{\text{PFR}}}{F_{\text{Ao}}} = \frac{1}{k}\left((1+\varepsilon_{\text{A}})\ln\left(\frac{1}{1-X_{\text{A}}}\right) - \varepsilon_{\text{A}}X_{\text{A}}\right)$$

Solving for the volume of the PFR gives

$$V_{\text{PFR}} = \frac{F_{\text{Ao}}}{kC_{\text{Ao}}}\left((1+\varepsilon_{\text{A}})\ln\left(\frac{1}{1-X_{\text{A}}}\right) - \varepsilon_{\text{A}}X_{\text{A}}\right)$$

Applying the preceding equation to the conditions of the problem statement, for a first-order reaction in a PFR, the reactor volume is

$$\begin{aligned} V_{\text{PFR}} &= \left(\frac{F}{k_2 C_{\text{M}}}\right)\left((1+\varepsilon)\ln\frac{1}{1-X} - \varepsilon X\right) \\ &= \left(\frac{25\ \frac{\text{lb mole}}{\text{min}}}{(0.363\ \text{min}^{-1})\left(0.0153\ \frac{\text{lb mole}}{\text{ft}^3}\right)}\right) \\ &\quad \times\left((1+2)\ln\frac{1}{1-0.85} - (2)(0.85)\right) \\ &= 17{,}966.47\ \text{ft}^3 \quad (1.8\times 10^4\ \text{ft}^3) \end{aligned}$$

The answer is (B).

62. The pressure of the water at the entrance of the valve, P_1, is 2200 lbf/in^2. The pressure of the water at the exit of the valve, P_2, is 1200 lbf/in^2. The difference between the pressure at the entrance and the exit of the valve is

$$\begin{aligned} \Delta P &= P_1 - P_2 \\ &= 2200\ \frac{\text{lbf}}{\text{in}^2} - 1200\ \frac{\text{lbf}}{\text{in}^2} \\ &= 1000\ \text{lbf/in}^2 \end{aligned}$$

The specific volume of the saturated water, v, is given as 0.0267 ft^3/lbm. From *NCEES Handbook* table "State Functions," the specific volume is

$$\nu = \frac{V}{m}$$

Solving the preceding equation for the mass gives

$$m = \frac{V}{\nu}$$

From *NCEES Handbook:* Density and Relative Density, the density is

$$\rho = \frac{m}{V} = \frac{\frac{V}{\nu}}{V} = \frac{1}{\nu}$$

At the valve conditions, the density of the saturated water is

$$\rho = \frac{1}{\nu} = \frac{1}{0.0267\ \frac{\text{ft}^3}{\text{lbm}}} = 37.5\ \text{lbm/ft}^3$$

From *NCEES Handbook:* Density, the specific gravity is

$$\text{SG} = \frac{\rho}{62.4\frac{\text{lbm}}{\text{ft}^3}}$$

The specific gravity of the saturated water is

$$\text{SG} = \frac{37.5\ \frac{\text{lbm}}{\text{ft}^3}}{62.4\ \frac{\text{lbm}}{\text{ft}^3}} = 0.601$$

It has been determined that when pure water at a temperature of 60°F flows through the valve at a flow rate of 425 gal/min, the pressure drop through the valve is 1 lbf/in^2. By definition, the flow coefficient of the control valve is

$$C_{\text{V}} = \frac{425\ \frac{\text{gal}}{\text{min}}}{\sqrt{1\ \frac{\text{lbf}}{\text{in}^2}}}$$

From *NCEES Handbook* table "Control Valve Sizing Equations for Liquids (Incompressible Flow)," the flow rate of the saturated water at the entrance of the valve is

$$\begin{aligned} \dot{V} &= C_{\text{V}}\sqrt{\frac{\Delta P}{\text{SG}}} \\ &= \frac{\left(\frac{425\ \frac{\text{gal}}{\text{min}}}{\sqrt{1\ \frac{\text{lbf}}{\text{in}^2}}}\right)\sqrt{\frac{1000\ \frac{\text{lbf}}{\text{in}^2}}{0.601}}}{7.48\ \frac{\text{gal}}{\text{ft}^3}} \\ &= 2317.66\ \text{ft}^3/\text{min} \end{aligned}$$

The mass flow rate of the water through the valve is

$$\begin{aligned}\dot{m}_w &= \dot{V}\rho \\ &= \left(2317.66\ \frac{\text{ft}^3}{\text{min}}\right)\left(37.5\ \frac{\text{lbm}}{\text{ft}^3}\right) \\ &= 86{,}912\ \text{lbm/min}\end{aligned}$$

At a pressure of 2200 lbf/in^2, the specific enthalpy of the saturated water, h_{1L}, is 695 Btu/lbm. At a pressure of 1200 lbf/in^2, the specific enthalpy of the liquid, h_L, is 572 Btu/lbm. At a pressure of 1200 lbf/in^2, the specific enthalpy of the vapor, h_v, is 1180 Btu/lbm. The liquid mass flow rate at the exit of the valve is

$$\begin{aligned}\dot{m}_L &= \dot{m}_w\left(\frac{h_{1L} - h_v}{h_L - h_v}\right) \\ &= \left(86{,}912\ \frac{\text{lbm}}{\text{min}}\right)\left(\frac{695\ \frac{\text{Btu}}{\text{lbm}} - 1180\ \frac{\text{Btu}}{\text{lbm}}}{572\ \frac{\text{Btu}}{\text{lbm}} - 1180\ \frac{\text{Btu}}{\text{lbm}}}\right) \\ &= 69{,}329\ \text{lbm/min}\end{aligned}$$

From *NCEES Handbook:* Material Balances, the general material balance equation is

$$\begin{aligned}\textit{Accumulation} &= \textit{Input} - \textit{Output} \\ &\quad + \textit{Generation} - \textit{Consumption}\end{aligned}$$

The vapor mass flow rate at the exit of the valve is

$$\begin{aligned}\dot{m}_v &= \dot{m}_w - \dot{m}_L \\ &= 86{,}912\ \frac{\text{lbm}}{\text{min}} - 69{,}329\ \frac{\text{lbm}}{\text{min}} \\ &= 17{,}583\ \text{lbm/min} \quad (18{,}000\ \text{lbm/min})\end{aligned}$$

The answer is (A).

63. The required removal efficiency, η, is 0.92. Because the vent stream consists of 37.5% toluene, the mole fraction of toluene in the vent stream, y, is 0.375. The total pressure, P, is given as 760 mm Hg. For the required removal efficiency, the partial pressure of toluene at the outlet of the condenser must be

$$\begin{aligned}P_{\text{toluene}} &= P\left(\frac{y(1-\eta)}{1-y\eta}\right) \\ &= (760\ \text{mm Hg})\left(\frac{(0.375)(1-0.92)}{1-(0.375)(0.92)}\right) \\ &= 34.81\ \text{mm Hg}\end{aligned}$$

At this pressure, the condensation temperature of toluene is

$$\begin{aligned}T_{\text{cond,°F}} &= \left(\frac{1344.8}{6.955 - \log P_{\text{toluene}}} - 219.48\right)(1.8°\text{F}) + 32°\text{F} \\ &= \left(\frac{1344.8}{6.955 - \log 34.81} - 219.48\right)(1.8°\text{F}) + 32°\text{F} \\ &= 84.10°\text{F}\end{aligned}$$

The volumetric flow rate of the vent stream is

$$\dot{Q} = \left(100\ \frac{\text{ft}^3}{\text{min}}\right)\left(60\ \frac{\text{min}}{\text{hr}}\right) = 6000\ \text{ft}^3/\text{hr}$$

The absolute temperature of the inlet gas stream is

$$T_{\text{in}} = 150°\text{F} + 460° = 610°\text{R}$$

From *NCEES Handbook:* Ideal Gas Law, the ideal gas law is

$$PV = nRT$$

Rearranging the preceding equation and solving for the molar volume gives

$$V = \frac{RT}{P}$$

Applying the preceding equation to the conditions of the problem statement, the molar specific volume occupied by the inlet gas stream at 610°R and 14.7 lbf/in^2 is

$$\begin{aligned}V_{\text{in}} &= \frac{R^* T_{\text{in}}}{P_{\text{in}}} \\ &= \frac{\left(1545\ \frac{\text{ft-lbf}}{\text{lb mole-°R}}\right)(610°\text{R})}{\left(14.7\ \frac{\text{lbf}}{\text{in}^2}\right)\left(12\ \frac{\text{in}}{\text{ft}}\right)^2} \\ &= 445\ \text{ft}^3/\text{lb mole}\end{aligned}$$

The molar flow rate of toluene in the inlet stream is

$$\begin{aligned}\dot{m}_{\text{in}} &= \left(\frac{\dot{Q}}{V_{\text{in}}}\right)y \\ &= \left(\frac{6000\ \frac{\text{ft}^3}{\text{hr}}}{445\ \frac{\text{ft}^3}{\text{lb mole}}}\right)(0.375) \\ &= 5.06\ \text{lb mole/hr}\end{aligned}$$

The molar flow rate of toluene in the gaseous outlet stream is

$$\begin{aligned}\dot{m}_{\text{out}} &= \dot{m}_{\text{in}}(1-\eta) \\ &= \left(5.06\ \frac{\text{lb mole}}{\text{hr}}\right)(1-0.92) \\ &= 0.405\ \text{lb mole/hr}\end{aligned}$$

From *NCEES Handbook:* Material Balances, the material balance general balance equation is

$$\begin{aligned}\textit{Accumulation} &= \textit{Input} - \textit{Output} \\ &\quad + \textit{Generation} - \textit{Consumption}\end{aligned}$$

Applying the preceding equation to the conditions of the problem statement, the molar flow rate of condensed toluene is

$$\begin{aligned}\dot{m}_{\text{cond}} &= \dot{m}_{\text{in}} - \dot{m}_{\text{out}} \\ &= 5.06\ \frac{\text{lb mole}}{\text{hr}} - 0.405\ \frac{\text{lb mole}}{\text{hr}} \\ &= 4.655\ \text{lb mole/hr}\end{aligned}$$

The average heat capacity of toluene, $c_{p,\text{toluene}}$, is given as 24.77 Btu/lb mole-°F. The average heat capacity of air, $c_{p,\text{air}}$, is 6.95 Btu/lb mole-°F. The heat of condensation of toluene, ΔH, is 14,395 Btu/lb mole. The temperature of the vent stream in degrees Fahrenheit, T_{in}, is 150°F. From *NCEES Handbook:* Heat Capacity/Specific Heat (c_p), the heat transferred is

$$\dot{Q} = \dot{m}c_p\Delta T$$

Applying the preceding equation to the conditions of the problem statement, the enthalpy change associated with the condensed toluene is

$$\begin{aligned}\Delta\dot{H}_{\text{cond}} &= \dot{m}_{\text{cond}}\big(\Delta H + c_{p,\text{toluene}}(T_{\text{in}} - T_{\text{cond}})\big) \\ &= \left(4.655\ \frac{\text{lb mole}}{\text{hr}}\right)\left(\begin{array}{l}14{,}395\ \dfrac{\text{Btu}}{\text{lb mole}} \\ +\left(24.77\ \dfrac{\text{Btu}}{\text{lb mole-°F}}\right) \\ \times(150\text{°F} - 84.10\text{°F})\end{array}\right) \\ &= 74{,}607\ \text{Btu/hr}\end{aligned}$$

The enthalpy change associated with the uncondensed toluene is

$$\begin{aligned}\Delta\dot{H}_{\text{uncond}} &= \dot{m}_{\text{out}}c_{p,\text{toluene}}(T_{\text{in}} - T_{\text{cond}}) \\ &= \left(0.405\ \frac{\text{lb mole}}{\text{hr}}\right)\left(24.77\ \frac{\text{Btu}}{\text{lb mole-°F}}\right) \\ &\quad \times(150\text{°F} - 84.10\text{°F}) \\ &= 661.099\ \text{Btu/hr}\end{aligned}$$

The enthalpy change associated with the noncondensible air is

$$\begin{aligned}\Delta\dot{H}_{\text{noncond}} &= \left(\frac{\dot{Q}}{V_{\text{in}}} - \dot{m}_{\text{in}}\right)c_{p,\text{air}}(T_{\text{in}} - T_{\text{cond}}) \\ &= \left(\frac{6000\ \frac{\text{ft}^3}{\text{hr}}}{445\ \frac{\text{ft}^3}{\text{lb mole}}} - 5.06\ \frac{\text{lb mole}}{\text{hr}}\right) \\ &\quad \times\left(6.95\ \frac{\text{Btu}}{\text{lb mole-°F}}\right)(150\text{°F} - 84.10\text{°F}) \\ &= 3857.843\ \text{Btu/hr}\end{aligned}$$

The condenser heat load is

$$\begin{aligned}\Delta\dot{H}_{\text{load}} &= \Delta\dot{H}_{\text{cond}} + \Delta\dot{H}_{\text{uncond}} + \Delta\dot{H}_{\text{noncond}} \\ &= 74{,}607\ \frac{\text{Btu}}{\text{hr}} + 661.099\ \frac{\text{Btu}}{\text{hr}} + 3857.843\ \frac{\text{Btu}}{\text{hr}} \\ &= 79{,}125.942\ \text{Btu/hr} \quad (7.9\times10^4\ \text{Btu/hr})\end{aligned}$$

The answer is (D).

64. From the requirements of room 1—a temperature of 24°C and relative humidity of 28%—and from *NCEES Handbook:* Psychrometric Chart (U.S. Customary Units), the specific volume of the air going to room 1, v_1, is 0.849 m^3/kg of dry air. From the requirements of room 2—a temperature of 32°C and relative humidity of 53%—and from the psychrometric chart, the specific volume of the air going to room 2, v_2, is 0.887 m^3/kg of dry air. The mass flow rate of the airstream going to room 1 is

$$\begin{aligned}\dot{m}_1 &= \frac{\dot{V}_1}{v_1} \\ &= \frac{134\ \frac{\text{m}^3}{\text{min}}}{0.849\ \frac{\text{m}^3}{\text{kg of dry air}}} \\ &= 157.8\ \text{kg of dry air/min}\end{aligned}$$

The mass flow rate of the airstream going to room 2 is

$$\begin{aligned}\dot{m}_2 &= \frac{\dot{V}_2}{v_2} \\ &= \frac{65\ \frac{\text{m}^3}{\text{min}}}{0.887\ \frac{\text{m}^3}{\text{kg of dry air}}} \\ &= 73.3 \text{ kg of dry air/min}\end{aligned}$$

From *NCEES Handbook:* Material Balances,

$$\begin{aligned}\textit{Accumulation} &= \textit{Input} - \textit{Output} \\ &\quad + \textit{Generation} - \textit{Consumption}\end{aligned}$$

This equation is used extensively throughout this problem solution.

The overall mass balance of dry air around the system gives the mass flow rate of the inlet fresh dry air.

$$\begin{aligned}\dot{m} &= \dot{m}_1 + \dot{m}_2 \\ &= 157.8\ \frac{\text{kg of dry air}}{\text{min}} + 73.3\ \frac{\text{kg of dry air}}{\text{min}} \\ &= 231.1 \text{ kg of dry air/min}\end{aligned}$$

At a dry-bulb temperature of 35°C and a wet-bulb temperature of 25°C (that is, the conditions of the fresh air) and from the psychrometric chart, the humidity ratio of the fresh air into the system, ω, is 0.0159 kg of water/kg of dry air. From the psychrometric chart, at the conditions of room 1, the humidity ratio of the air into room 1, ω_1, is 0.005 17 kg of water/kg of dry air. From the psychrometric chart, at the conditions of room 2, the humidity ratio of the air into room 2, ω_2, is 0.0159 kg of water/kg of dry air. The mass flow rate of water in the fresh air is

$$\begin{aligned}\dot{m}_w &= \dot{m}\omega \\ &= \left(231.1\ \frac{\text{kg of dry air}}{\text{min}}\right)\left(0.0159\ \frac{\text{kg of water}}{\text{kg of dry air}}\right) \\ &= 3.674 \text{ kg of water/min}\end{aligned}$$

The mass flow rate of water in the air going into room 1 is

$$\begin{aligned}\dot{m}_{w1} &= \dot{m}_1\omega_1 \\ &= \left(157.8\ \frac{\text{kg of dry air}}{\text{min}}\right)\left(0.005\,17\ \frac{\text{kg of water}}{\text{kg of dry air}}\right) \\ &= 0.816 \text{ kg of water/min}\end{aligned}$$

The mass flow rate of water in the air going into room 2 is

$$\begin{aligned}\dot{m}_{w2} &= \dot{m}_2\omega_2 \\ &= \left(73.3\ \frac{\text{kg of dry air}}{\text{min}}\right)\left(0.0159\ \frac{\text{kg of water}}{\text{kg of dry air}}\right) \\ &= 1.165 \text{ kg of water/min}\end{aligned}$$

An overall mass balance of water around the system gives the mass flow rate of the water out of the cooler/condenser.

$$\begin{aligned}\dot{m}_{w3} &= \dot{m}_w - \dot{m}_{w1} - \dot{m}_{w2} \\ &= 3.674\ \frac{\text{kg of water}}{\text{min}} - 0.816\ \frac{\text{kg of water}}{\text{min}} \\ &\quad -1.165\ \frac{\text{kg of water}}{\text{min}} \\ &= 1.69 \text{ kg of water/min} \quad (1.7 \text{ kg/min})\end{aligned}$$

The answer is (C).

65. The mass flow rate is

$$\dot{m} = \frac{27\ \frac{\text{kg}}{\text{min}}}{60\ \frac{\text{s}}{\text{min}}} = 0.45 \text{ kg/s}$$

The average heat capacity of the water in the pipe, c_p, is given as 4280 J/kg·K. The diameter of the pipe is

$$D = \frac{50 \text{ mm}}{1000\ \frac{\text{mm}}{\text{m}}} = 0.05 \text{ m}$$

Because steam is condensing on the outer surface of the pipe, from *NCEES Handbook:* Log-Mean Temperature Difference, the logarithmic mean temperature difference is

$$\begin{aligned}\Delta T_{\text{lm}} &= \frac{(T_{\text{hot,out}} - T_{\text{cold,in}}) - (T_{\text{hot,in}} - T_{\text{cold,out}})}{\ln\left(\frac{T_{\text{hot,out}} - T_{\text{cold,in}}}{T_{\text{hot,in}} - T_{\text{cold,out}}}\right)} \\ &= \frac{(90°\text{C} - 20°\text{C}) - (90°\text{C} - 68°\text{C})}{\ln\left(\frac{90°\text{C} - 20°\text{C}}{90°\text{C} - 68°\text{C}}\right)} \\ &= 41.47°\text{C}\end{aligned}$$

A temperature difference of 1°C is equal to 1K, so

$$\Delta T_{\text{lm}} = 41.47\text{K}$$

From *NCEES Handbook:* Heat Capacity/Specific Heat (c_p), the heat transferred in and out of a flowing material is

$$\dot{Q} = \dot{m}c_p\Delta T$$

Condensation occurs at a constant temperature, so the heat transfer is

$$\begin{aligned}\overline{Q} &= \dot{m}c_p(T_{\text{cold,out}} - T_{\text{cold,in}})\\ &= \left(0.45\ \frac{\text{kg}}{\text{s}}\right)\left(4280\ \frac{\text{J}}{\text{kg·K}}\right)(68°\text{C} - 20°\text{C})\\ &= 92\,448\ \text{W}\end{aligned}$$

From *NCEES Handbook:* Newton's Law of Cooling, the average convection coefficient can be found from the formula

$$\begin{aligned}\dot{Q} &= hA\Delta T_{\text{lm}}\\ &= h(\pi DL)\Delta T_{\text{lm}}\\ h &= \frac{\dot{Q}}{\pi DL\Delta T_{\text{lm}}}\\ &= \frac{92\,448\ \text{W}}{\pi(0.05\ \text{m})(10\ \text{m})(41.47\text{K})}\\ &= 1419.2\ \text{W/m}^2\text{·K}\quad (1400\ \text{W/m}^2\text{·K})\end{aligned}$$

The answer is (B).

66. The heat capacity of n-butane, $c_{p,n\text{-butane}}$, is 141 J/mol·K. The heat capacity of the inert material, $c_{p,\text{inert}}$, is 161 J/mol·K. The feed contains 86% n-butane and 14% inert ingredients, so the heat capacity of the feed is

$$\begin{aligned}c_{p,\text{feed}} &= c_{p,n\text{-butane}} + \left(\frac{0.14}{0.86}\right)c_{p,\text{inert}}\\ &= 141\ \frac{\text{J}}{\text{mol·K}} + \left(\frac{0.14}{0.86}\right)\left(161\ \frac{\text{J}}{\text{mol·K}}\right)\\ &= 167.2\ \text{J/mol·K}\end{aligned}$$

The temperature of the feed, T_{feed}, is 345K. The heat of reaction $\widehat{h}_r$, is 6900 J/mol. The conversion, X, is 0.55. Because the reactor is adiabatic and the reaction is exothermic, the heat produced by the reaction increases the temperature of the components in the reactor. An energy balance around the reactor gives

$$c_{p,\text{feed}}(T_{\text{product}} - T_{\text{feed}}) = \Delta\widehat{h}_r X$$

Solving for the temperature of the reaction mixture gives

$$\begin{aligned}T_{\text{product}} &= T_{\text{feed}} + \left(\frac{\Delta\widehat{h}_r}{c_{p,\text{feed}}}\right)X\\ &= 345\text{K} + \left(\frac{6900\ \frac{\text{J}}{\text{mol}}}{167.2\ \frac{\text{J}}{\text{mol·K}}}\right)(0.55)\\ &= 368\text{K}\end{aligned}$$

The reaction rate constant, k_2, is 62.2 h^{-1}. The temperature at which the rate reaction constant has been determined, T_1, is 380K. The universal gas constant, R, is 8.31 J/mol·K. The activation energy is

$$E_a = \left(65.7\ \frac{\text{kJ}}{\text{mol}}\right)\left(1000\ \frac{\text{J}}{\text{kJ}}\right) = 65\,700\ \text{J/mol}$$

From *NCEES Handbook:* Temperature Dependence, the activation energy is

$$\ln\left(\frac{k_1}{k_2}\right) = \frac{E_a}{R}\,\frac{T_1 - T_2}{T_1T_2}$$

The reaction rate constant at the reaction temperature of 368K is

$$\ln\left(\frac{k_1}{k_2}\right) = \frac{E_a}{R}\,\frac{T_{\text{product}} - T_1}{T_{\text{product}}T_1}$$

Solving for k_1 gives

$$\begin{aligned}k_1 &= k_2 e^{(E_a/R)\left((1/T_1)-(1/T_{\text{product}})\right)}\\ &= (62.2\ \text{h}^{-1})e^{\left(65\,700\ \frac{\text{J}}{\text{mol}}\Big/ 8.31\ \frac{\text{J}}{\text{mol·K}}\right)(1/380\text{K}-1/368\text{K})}\\ &= 31.56\ \text{h}^{-1}\end{aligned}$$

The equilibrium constant, K_1, is 3.09. The temperature at which the equilibrium constant has been determined, T_{K_1}, is 335K. From *NCEES Handbook:* Effect of Temperature on Chemical Equilibrium Constants, the equilibrium constant at the reaction temperature of 368K is

$$\ln\frac{K_2}{K_1} = \left(\frac{\Delta\widehat{h}_r}{R}\right)\left(\frac{1}{T_{K_1}} - \frac{1}{T}\right)$$

Solving for K_2 gives

$$K_2 = K_1 e^{(-\Delta\hat{h}_r/R)\left((1/T_{K_1})-(1/T_{\text{product}})\right)}$$
$$= 3.09 e^{\left(-6900\ \frac{\text{J}}{\text{mol}} \Big/ 8.31\ \frac{\text{J}}{\text{mol·K}}\right)(1/335\text{K}-1/368\text{K})}$$
$$= 2.474$$

Let the rate constant of the forward reaction, $n\text{-}C_4H_{10} \rightarrow i\text{-}C_4H_{10}$, be k_1. Let the rate constant of the inverse reaction, $i\text{-}C_4H_{10} \rightarrow n\text{-}C_4H_{10}$, be k_2. From *NCEES Handbook:* First-Order Reversible Reactions, the rate of reaction is

$$-r_A = k_1 C_A - k_2 C_R$$

The rate of reaction is the rate of the forward reaction minus the rate of the inverse reaction.

$$-r = k_1 C_0 (1 - X) - k_2 C_0 X$$

From *NCEES Handbook:* Liquid Phase Reactions, the rate constant is

$$K_c = \frac{k_1}{k_2}$$

Replacing k_2 with k_1/K in the rate constant equation gives

$$-r = k_1 C_0(1-X) - \left(\frac{k_1}{K_2}\right) C_0 X$$
$$= k_1 C_0 \left(1 - \left(1 + \frac{1}{K_2}\right) X\right)$$
$$= (31.56\ \text{h}^{-1})\left(11.87\ \frac{\text{kmol}}{\text{m}^3}\right)\left(1 - \left(1 + \frac{1}{2.474}\right)(0.55)\right)$$
$$= 85.30\ \text{kmol/m}^3\text{·h}$$

The feed, F, is 400 kmol/h. The feed is 86% n-butane, so the molar feed rate of n-butane is

$$F_{C_4H_{10}} = 0.86F$$
$$= (0.86)\left(400\ \frac{\text{kmol}}{\text{h}}\right)$$
$$= 344\ \text{kmol/h}$$

The volume of the reactor is

$$V = \left(\frac{F_{C_4H_{10}}}{-r}\right) X$$
$$= \left(\frac{344\ \frac{\text{kmol}}{\text{h}}}{85.30\ \frac{\text{kmol}}{\text{m}^3\text{·h}}}\right)(0.55)$$
$$= 2.218\ \text{m}^3 \quad (2.2\ \text{m}^3)$$

The answer is (C).

67. The absolute temperatures of the air and the inner and outer surfaces are

$$T_{\text{air}} = 1500°\text{C} + 273° = 1773\text{K}$$
$$T_{\text{inner}} = 860°\text{C} + 273° = 1133\text{K}$$
$$T_{\text{outer}} = 20°\text{C} + 273° = 293\text{K}$$

The inside convection coefficient, h, is given as 25 W/m^2·K. From *NCEES Handbook:* Newton's Law of Cooling, the heat flow rate between the oven air and the inner surface is

$$\dot{Q} = hA(T_w - T_\infty)$$
$$\frac{\dot{Q}}{A_s} = h(T_{\text{air}} - T_{\text{inner}})$$
$$\dot{q}_1 = h(T_{\text{air}} - T_{\text{inner}})$$
$$= \left(25\ \frac{\text{W}}{\text{m}^2\text{·K}}\right)(1773\text{K} - 1133\text{K})$$
$$= 16\,000\ \text{W/m}^2$$

The refractory brick's thermal resistance is

$$\frac{\delta_r}{k_r} = \frac{0.30\ \text{m}}{45\ \frac{\text{W}}{\text{m·K}}} = 0.00667\ \text{m}^2\text{·K/W}$$

The insulation brick's thermal resistance to heat transfer is

$$\frac{\delta_i}{k_i} = \frac{0.20\ \text{m}}{20\ \frac{\text{W}}{\text{m·K}}} = 0.01\ \text{m}^2\text{·K/W}$$

The firebrick's thermal resistance to heat transfer is

$$\frac{\delta_f}{k_f} = \frac{0.15\ \text{m}}{k_f}$$

From *NCEES Handbook:* Conduction, the heat flow rate between the inner and outer surfaces is

$$\dot{Q} = \frac{A(T_1 - T_2)}{\sum_i \frac{\delta_i}{k_i}}$$

$$\frac{\dot{Q}}{A} = \dot{q} = \frac{T_{\text{inner}} - T_{\text{outer}}}{\frac{\delta_r}{k_r} + \frac{\delta_f}{k_f} + \frac{\delta_i}{k_i}}$$

The heat is transferring under steady-state conditions, so the heat flow rate between the oven air and the inner surface equals the heat flow rate between the inner and outer surfaces.

$$\dot{q}_1 = \dot{q}_2 = \frac{T_{\text{inner}} - T_{\text{outer}}}{\frac{\delta_r}{k_r} + \frac{\delta_f}{k_f} + \frac{\delta_i}{k_i}}$$

Solving for the thermal conductivity of the firebrick,

$$\begin{aligned} k_f &= \frac{\delta_f}{\frac{T_{\text{inner}} - T_{\text{outer}}}{\dot{q}_1} - \frac{\delta_r}{k_r} - \frac{\delta_i}{k_i}} \\ &= \frac{0.15 \text{ m}}{\frac{1133\text{K} - 293\text{K}}{16\,000 \frac{\text{W}}{\text{m}^2}} - 0.00667 \frac{\text{m}^2\cdot\text{K}}{\text{W}} - 0.01 \frac{\text{m}^2\cdot\text{K}}{\text{W}}} \\ &= 4.186 \text{ W/m}\cdot\text{K} \quad (4.2 \text{ W/m}\cdot\text{K}) \end{aligned}$$

The answer is (C).

68. Throughout this solution, the balance equation from *NCEES Handbook:* Material Balances is used repeatedly. To find the number of moles of flue gas that are produced with every 1 lb mole of feed, calculate a carbon mass balance around the combustion chamber. From the given composition of the combustion gas, 1 lb mole of feed contains

component	mass (lb mole)
C_2H_4	0.1510
C_3H_8	0.1420
CO_2	0.2810
CO	0.0290

From the given composition of the flue gas, 1 lb mole of flue gas contains

component	mass (lb mole)
CO_2	0.1150
CO	0.0035

A carbon mole balance around the combustion chamber gives

$$\begin{aligned} n_{\text{feed}} &= \frac{n_{\text{C,feed}}}{n_{\text{C,flue}}} \\ &= \frac{2n_{\text{C}_2\text{H}_4\text{,feed}} + 3n_{\text{C}_3\text{H}_8\text{,feed}} + n_{\text{CO}_2\text{,feed}} + n_{\text{CO,feed}}}{n_{\text{CO}_2\text{,flue}} + n_{\text{CO,flue}}} \\ &= \frac{\begin{array}{c}(2)(0.1510 \text{ lb mole}) + (3)(0.1420 \text{ lb mole}) \\ +0.2810 \text{ lb mole} + 0.0290 \text{ lb mole}\end{array}}{0.1150 \text{ lb mole} + 0.0035 \text{ lb mole}} \\ &= 8.759 \text{ lb mole of flue gas/lb mole of feed} \end{aligned}$$

To balance the carbon mass, then, there is 8.759 lb mole of flue gas produced for every 1.0 lb mole of feed.

Nitrogen does not react during the combustion, so the number of moles of nitrogen in the feed and air combined equals the number of moles of nitrogen in the flue gas. From the problem statement, nitrogen is 4.30% of the feed, so there is 0.0430 lb mole of nitrogen for every 1.0 lb mole of feed.

Also from the problem statement, nitrogen is 82.90% of the flue gas, so for every 1.0 lb mole of feed, the amount of nitrogen in the flue gas is

$$n_{\text{N}_2\text{,flue}} = (8.759 \text{ lb mole})(0.8290) = 7.261 \text{ lb mole}$$

Of this nitrogen, 0.0430 lb mole comes from the feed, so the rest must come from the air. The amount of nitrogen in the flue gas that comes from the air, then, is

$$\begin{aligned} n_{\text{N}_2\text{,air}} &= n_{\text{N}_2\text{,flue}} - n_{\text{N}_2\text{,feed}} \\ &= 7.261 \text{ lb mole} - 0.0430 \text{ lb mole} \\ &= 7.218 \text{ lb mole} \end{aligned}$$

Air is 79% nitrogen, so the total amount of air must be

$$\begin{aligned} n_{\text{air}} &= \frac{n_{\text{N}_2\text{,air}}}{0.79} = \frac{7.218 \text{ lb mole}}{0.79} \\ &= 9.137 \text{ lb mole} \quad (9 \text{ lb mole}) \end{aligned}$$

The answer is (D).

69. The absolute temperature of the wall's outer surface is

$$T_{\text{w}} = 126^\circ\text{C} + 273^\circ = 399\text{K}$$

The absolute temperature of the air is

$$T_\infty = 23^\circ\text{C} + 273^\circ = 296\text{K}$$

The heat is transferred from the wall's inner surface, through the wall by conduction, to the wall's outer surface. From there, the heat is transferred to the surrounding air by convection and radiation. The heat at the inner surface can be determined by following the path of the heat backward from the surrounding air. From *NCEES Handbook:* Newton's Law of Cooling, the heat transfer rate by convection is

$$\dot{Q} = hA(T_\text{w} - T_\infty)$$

Per unit area, the heat transfer rate by convection is

$$\begin{aligned}\dot{Q}_\text{conv} &= h(T_\text{w} - T_\infty)\\ &= \left(25\ \frac{\text{W}}{\text{m}^2\cdot\text{K}}\right)(399\text{K} - 296\text{K})\\ &= 2575\ \text{W/m}^2\end{aligned}$$

The emissivity, ε, is given as 0.85. The Stefan-Boltzmann constant, σ, is $5.67\times10^{-8}\ \text{W/m}^2\cdot\text{K}^4$. From *NCEES Handbook:* Stefan-Boltzmann Law of Radiation, the heat transfer rate by radiation is

$$\dot{Q} = \varepsilon\sigma A T^4$$

The heat transfer rate by radiation, per unit area, is

$$\begin{aligned}\dot{Q}_\text{rad} &= \varepsilon\sigma(T_\text{w}^4 - T_\infty^4)\\ &= (0.85)\left(5.67\times10^{-8}\ \frac{\text{W}}{\text{m}^2\cdot\text{K}^4}\right)\left((399\text{K})^4 - (296\text{K})^4\right)\\ &= 851.5\ \text{W/m}^2\end{aligned}$$

When the wall is at steady-state conditions, because of the conservation of energy, the heat rate transfer by conduction equals the heat rate transfer by convection plus the heat rate transfer by radiation. The heat rate transfer by conduction is

$$\dot{Q}_\text{cond} = \frac{k(T_\text{inner} - T_\text{w})}{L}$$

The energy balance at the outer surface of the wall is

$$\dot{Q}_\text{cond} = \dot{Q}_\text{conv} + \dot{Q}_\text{rad}$$

Combining these two equations gives

$$q_\text{conv} + q_\text{rad} = \frac{k(T_\text{inner} - T_\text{outer})}{L}$$

Solving for T_inner in degrees Celsius gives

$$\begin{aligned}T_\text{inner} &= \frac{(\dot{Q}_\text{conv} + \dot{Q}_\text{rad})L}{k} + T_\text{w}\\ &= \frac{\left(2575\ \frac{\text{W}}{\text{m}^2} + 851.5\ \frac{\text{W}}{\text{m}^2}\right)(0.2\ \text{m})}{1.6\ \frac{\text{W}}{\text{m}\cdot\text{K}}} + 399\text{K}\\ &= 827.3\text{K} - 273^\circ\\ &= 554.3^\circ\text{C}\quad(550^\circ\text{C})\end{aligned}$$

The answer is (C).

70. The plant will operate 300 day/yr. The feed to the evaporator system, $\dot{m}_\text{in}$, is 900,000 lbm/day. The number of years of service life, n, is 10 yr. The salvage value for each effect, S, is \$10,000. The initial investment per effect, P, is \$36,000. The number of effects, as yet undetermined, is N.

The total cost per day is the sum of the depreciation per operating day, D, the fixed costs per day, C_fixed, the steam cost per day, C_steam, and the remaining costs per day, C_rem.

$$C_\text{total} = D + C_\text{fixed} + C_\text{steam} + C_\text{rem}$$

As in *NCEES Handbook* table "Depreciation Methods," the depreciation per operating day, using the straight-line method, is

$$\begin{aligned}D &= \frac{N(P-S)}{nt} = \frac{N(\$36{,}000 - \$10{,}000)}{(10\ \text{yr})\left(300\ \frac{\text{day}}{\text{yr}}\right)}\\ &= N\left(\frac{\$8.67}{\text{day}}\right)\end{aligned}$$

The annual fixed costs other than depreciation will equal 20% of the initial investment, so these costs per operating day are

$$\begin{aligned}C_\text{fixed} &= \frac{NP\left(\frac{20\%}{\text{yr}}\right)}{t} = \frac{N(\$36{,}000)\left(\frac{0.20}{\text{yr}}\right)}{300\ \frac{\text{day}}{\text{yr}}}\\ &= N\left(\frac{\$24.00}{\text{day}}\right)\end{aligned}$$

The liquor in the feed contains 6% caustic soda by weight, so the mass fraction of soda in the feed, w_in, is 0.06. The product contains 38% caustic soda by weight, so the mass fraction of soda in the product, w_{out}, is 0.38. The amount of soda in the feed, then, is

$$\begin{aligned}\dot{m}_{soda} &= \dot{m}_{in}w_{in}\\ &= \left(900{,}000\ \frac{lbm}{day}\right)(0.06)\\ &= 54{,}000\ lbm/day\end{aligned}$$

The same amount of soda is in the product, so

$$\dot{m}_{soda} = \dot{m}_{out}w_{out}$$

The total amount of product is

$$\begin{aligned}\dot{m}_{out} &= \frac{\dot{m}_{soda}}{w_{out}} = \frac{54{,}000\ \frac{lbm}{day}}{0.38}\\ &= 142{,}100\ lbm/day\end{aligned}$$

The amount of water that must be removed by evaporation is

$$\begin{aligned}\dot{m}_{evap} &= \dot{m}_{in} - \dot{m}_{out}\\ &= 900{,}000\ \frac{lbm}{day} - 142{,}100\ \frac{lbm}{day}\\ &= 757{,}900\ lbm/day\end{aligned}$$

The amount of steam required is given as 1.11 times the water evaporated in each effect. This is

$$\begin{aligned}\dot{m}_{steam} &= 1.11\left(\frac{\dot{m}_{evap}}{N}\right) = (1.11)\left(\frac{757{,}900\ \frac{lbm}{day}}{N}\right)\\ &= \frac{841{,}300\ \frac{lbm}{day}}{N}\end{aligned}$$

The cost of the steam is given as $0.72 per 1000 lbm. The daily cost of the steam to run the plant is

$$\begin{aligned}C_{steam} &= \left(\frac{\$0.72}{1000\ lbm}\right)\left(\frac{841{,}300\ \frac{lbm}{day}}{N}\right)\\ &= \left(\frac{1}{N}\right)\left(\frac{\$605.7}{day}\right)\end{aligned}$$

Regardless of the number of effects, the remaining costs are

$$C_{rem} = \$80/day$$

The total cost is

$$\begin{aligned}C_{total} &= D + C_{fixed} + C_{steam} + C_{rem}\\ &= N\left(\frac{\$8.67}{day}\right) + N\left(\frac{\$24.00}{day}\right) + \left(\frac{1}{N}\right)\left(\frac{\$605.7}{day}\right) + \frac{\$80}{day}\\ &= \frac{N(\$32.67) + \frac{\$605.7}{N} + \$80}{day}\end{aligned}$$

For each number of effects, the total cost is

$$C_{1,total} = \frac{(1)(\$32.67) + \frac{\$605.7}{1} + \$80}{day} = \$718.37/day$$

$$C_{2,total} = \frac{(2)(\$32.67) + \frac{\$605.7}{2} + \$80}{day} = \$448.19/day$$

$$C_{3,total} = \frac{(3)(\$32.67) + \frac{\$605.7}{3} + \$80}{day} = \$379.91/day$$

$$C_{4,total} = \frac{(4)(\$32.67) + \frac{\$605.7}{4} + \$80}{day} = \$362.10/day$$

$$C_{5,total} = \frac{(5)(\$32.67) + \frac{\$605.7}{5} + \$80}{day} = \$364.49/day$$

Each added effect divides the daily steam cost further, but the savings in steam cost from each added effect become less as the number of effects grows. At five effects, the savings become less than the $32.67/day that each effect adds to the depreciation and fixed charges, so adding more effects after five will only increase the total daily cost. The four-effect (4) evaporator has the lowest daily cost.

The answer is (C).

71. The relevant data are shown in the illustration.

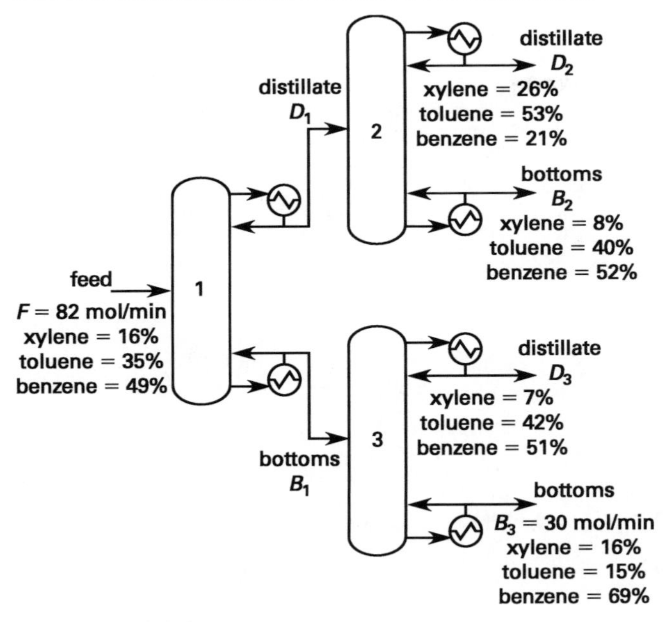

From *NCEES Handbook:* Material Balances,

$$\textit{Accumulation} = \textit{Input} - \textit{Output} + \textit{Generation} - \textit{Consumption}$$

is used repeatedly throughout this solution. The mass balance of xylene around the entire system gives

$$\begin{aligned}
&0.26D_2 + 0.08B_2 + 0.07D_3 + 0.16B_3 \\
&\quad = 0.16F \\
&0.26D_2 + 0.08B_2 + 0.07D_3 \\
&\quad = 0.16F - 0.16B_3 \\
&\quad = (0.16)\left(82\ \frac{\text{mol}}{\text{min}}\right) - (0.16)\left(30\ \frac{\text{mol}}{\text{min}}\right) \\
&\quad = 8.32\ \text{mol/min}
\end{aligned}$$

The mass balance of toluene around the entire system gives

$$\begin{aligned}
&0.53D_2 + 0.40B_2 + 0.42D_3 + 0.15B_3 \\
&\quad = 0.35F \\
&0.53D_2 + 0.40B_2 + 0.42D_3 \\
&\quad = 0.35F - 0.15B_3 \\
&\quad = (0.35)\left(82\ \frac{\text{mol}}{\text{min}}\right) - (0.15)\left(30\ \frac{\text{mol}}{\text{min}}\right) \\
&\quad = 24.2\ \text{mol/min}
\end{aligned}$$

The mass balance of benzene around the entire system gives

$$0.21D_2 + 0.52B_2 + 0.51D_3 + 0.69B_3 = 0.49F$$

$$\begin{aligned}
&0.21D_2 + 0.52B_2 + 0.51D_3 \\
&\quad = 0.49F - 0.69B_3 \\
&\quad = (0.49)\left(82\ \frac{\text{mol}}{\text{min}}\right) - (0.69)\left(30\ \frac{\text{mol}}{\text{min}}\right) \\
&\quad = 19.48\ \text{mol/min}
\end{aligned}$$

Therefore, the three mass balance equations are

$$\begin{aligned}
0.26D_2 + 0.08B_2 + 0.07D_3 &= 8.32\ \text{mol/min} \\
0.53D_2 + 0.40B_2 + 0.42D_3 &= 24.2\ \text{mol/min} \\
0.21D_2 + 0.52B_2 + 0.51D_3 &= 19.48\ \text{mol/min}
\end{aligned}$$

Solving these equations simultaneously gives

$$\begin{aligned}
D_2 &= 23.92\ \text{mol/min} \\
B_2 &= 13.55\ \text{mol/min} \\
D_3 &= 14.53\ \text{mol/min}
\end{aligned}$$

An overall mass balance around distillation column 3 gives

$$B_1 = D_3 + B_3 = 14.53\ \frac{\text{mol}}{\text{min}} + 30\ \frac{\text{mol}}{\text{min}} = 44.53\ \text{mol/min}$$

The mass balance of benzene around distillation column 3 gives

$$\begin{aligned}
B_{1,\text{benzene}} &= 0.51D_3 + 0.69B_3 \\
&= (0.51)\left(14.53\ \frac{\text{mol}}{\text{min}}\right) + (0.69)\left(30\ \frac{\text{mol}}{\text{min}}\right) \\
&= 28.11\ \text{mol/min}
\end{aligned}$$

The mole fraction of benzene in the bottoms stream from column 1 is

$$\begin{aligned}
x &= \frac{B_{1,\text{benzene}}}{B_1} = \frac{28.11\ \frac{\text{mol}}{\text{min}}}{44.53\ \frac{\text{mol}}{\text{min}}} \\
&= 0.6313 \quad (0.63)
\end{aligned}$$

The answer is (C).

72. The relevant data are shown in the illustration.

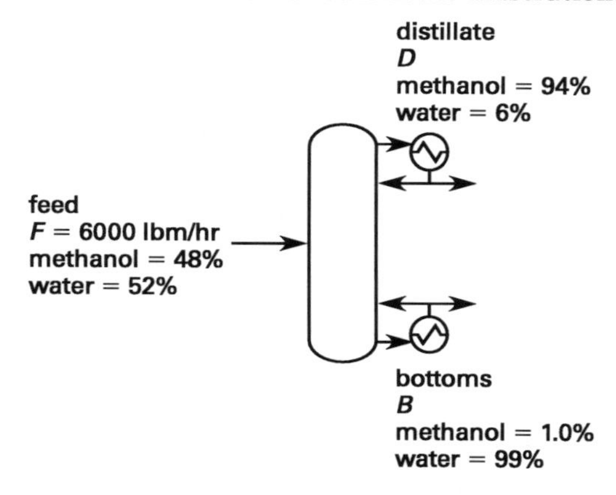

The weight fraction of methanol in the feed, w_F, is 0.48. The weight fraction of methanol in the distillate, w_D, is 0.94. The weight fraction of methanol in the bottom, w_B, is 0.01. From *NCEES Handbook:* Column Material Balance, the overall material balance is

$$F = B + D$$

Rearranging, an overall mass balance around the distillation column gives

$$B = F - D$$

A mass balance of methanol around the distillation column gives

$$w_F F = w_D D + w_B B$$

Substituting and solving for the mass flow rate of the distillate produced gives

$$\begin{aligned} w_F F &= w_D D + w_B(F - D) \\ &= w_D D + w_B F - w_B D \\ w_F F - w_B F &= w_D D - w_B D \\ F(w_F - w_B) &= D(w_D - w_B) \end{aligned}$$

$$\begin{aligned} D &= \frac{F(w_F - w_B)}{w_D - w_B} = \frac{\left(6000\ \frac{\text{lbm}}{\text{hr}}\right)(0.48 - 0.01)}{0.94 - 0.01} \\ &= 3032\ \text{lbm/hr} \quad (3000\ \text{lbm/hr}) \end{aligned}$$

The answer is (B).

73. The relevant data are shown in the illustration.

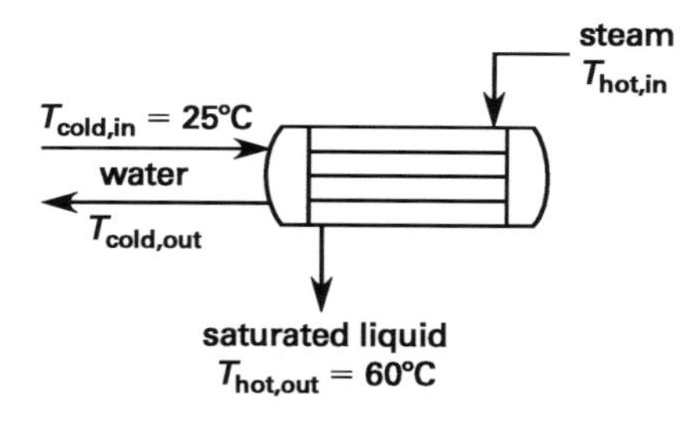

The diameter of the tubes is

$$D = \frac{50\ \text{mm}}{1000\ \frac{\text{mm}}{\text{m}}} = 0.05\ \text{m}$$

The number of tubes, N, is given as 10 000. The mass flow rate of the cooling water, $\dot{m}_w$, is given as 36 000 kg/s. The mass flow rate per tube is

$$\dot{m} = \frac{\dot{m}_w}{N} = \frac{36\,000\ \frac{\text{kg}}{\text{s}}}{10\,000} = 3.6\ \text{kg/s}$$

The rate of heat transferred, Q, is given as 8.53×10^8 W. The heat capacity of the cooling water, c_p, is given as 4183 J/kg·K. From *NCEES Handbook:* Heat Capacity/Specific Heat (c_p), the heat transferred is

$$\dot{Q} = \dot{m} c_p \Delta T$$

In terms of the temperature difference, the preceding equation gives

$$\dot{Q} = \dot{m} c_p (T_{\text{cold,out}} - T_{\text{cold,in}})$$

An overall energy balance gives the water outlet temperature. (A temperature increase of 1K is equal to a temperature increase of 1°C, so no conversion is needed.)

$$\begin{aligned} T_{\text{cold,out}} &= T_{\text{cold,in}} + \frac{\dot{Q}}{\dot{m}_w c_p} \\ &= 25°\text{C} + \frac{8.53 \times 10^8\ \text{W}}{\left(36\,000\ \frac{\text{kg}}{\text{s}}\right)\left(4183\ \frac{\text{J}}{\text{kg·K}}\right)} \\ &= 30.66°\text{C} \end{aligned}$$

The steam-side convection coefficient, h_o, is given as 1223 W/m²·K. The thermal conductivity of the cooling water, k, is given as 0.61 W/m·K. The viscosity of the cooling water, μ, is given as 0.0009 N·s/m². From

NCEES Handbook table "Dimensionless Numbers," the Reynolds number is

$$\begin{aligned} Re &= \frac{\rho u D}{\mu} = \frac{D\rho\left(\frac{\dot{Q}}{A}\right)}{\mu} = \frac{\rho D\dot{Q}}{A\mu} \\ &= \frac{\rho D\dot{Q}}{\left(\frac{\pi D^2}{4}\right)\mu} \\ &= \frac{4\dot{m}}{\pi D\mu} \\ &= \frac{(4)\left(3.6\ \frac{\text{kg}}{\text{s}}\right)}{\pi(0.05\ \text{m})\left(0.0009\ \frac{\text{N}\cdot\text{s}}{\text{m}^2}\right)} \\ &= 101\,859 \end{aligned}$$

The Prandtl number, Pr, is given as 5.90. From the Dittus-Boelter correlation, the average Nusselt number is

$$\begin{aligned} \overline{Nu} &= 0.023(Re)^{4/5}(Pr)^{0.4} \\ &= (0.023)(101\,859)^{4/5}(5.90)^{0.4} \\ &= 474.75 \end{aligned}$$

The tube-side convection coefficient is

$$\begin{aligned} h_i &= (\overline{Nu})\left(\frac{k}{D}\right) \\ &= (474.75)\left(\frac{0.61\ \frac{\text{W}}{\text{m}\cdot\text{K}}}{0.05\ \text{m}}\right) \\ &= 5792\ \text{W/m}^2\cdot\text{K} \end{aligned}$$

From *NCEES Handbook:* Overall Heat-Transfer Coefficient, the overall heat transfer coefficient is

$$\begin{aligned} \frac{1}{U_{\text{ov}}A_{\text{ref}}} &= \frac{1}{h_{\text{i}}A_{\text{i}}} + \frac{R_{\text{fi}}}{A_{\text{i}}} + \frac{\ln\left(\frac{D_o}{D_i}\right)}{2\pi kL} \\ &\quad + \frac{R_{\text{fo}}}{A_{\text{o}}} + \frac{1}{h_{\text{o}}A_{\text{o}}} \end{aligned}$$

Each tube is of thin-wall construction, so the resistance to heat transfer through the wall of the tube may be considered negligible. With fouling factor for inside and outside the tubes zero, the overall heat transfer coefficient is

$$\begin{aligned} U_{\text{ov}} &= \frac{1}{\frac{1}{h_{\text{i}}} + \frac{1}{h_{\text{o}}}} = \frac{1}{\frac{1}{5792\ \frac{\text{W}}{\text{m}^2\cdot\text{K}}} + \frac{1}{1223\ \frac{\text{W}}{\text{m}^2\cdot\text{K}}}} \\ &= 1010\ \text{W/m}^2\cdot\text{K} \end{aligned}$$

The steam condenses on the outer surface of the tubes at a given temperature of

$$T_{\text{hot,in}} = 60^\circ\text{C}$$

From *NCEES Handbook:* Log-Mean Temperature Difference, the logarithmic mean temperature difference (LMTD) is

$$\begin{aligned} \Delta T_{\text{lm}} &= \frac{(T_{\text{hot,out}} - T_{\text{cold,in}}) - (T_{\text{hot,in}} - T_{\text{cold,out}})}{\ln\left(\frac{T_{\text{hot,out}} - T_{\text{cold,in}}}{T_{\text{hot,in}} - T_{\text{cold,out}}}\right)} \\ &= \frac{(60^\circ\text{C} - 25^\circ\text{C}) - (60^\circ\text{C} - 30.66^\circ\text{C})}{\ln\left(\frac{60^\circ\text{C} - 25^\circ\text{C}}{60^\circ\text{C} - 30.66^\circ\text{C}}\right)} \\ &= 32.087^\circ\text{C} \end{aligned}$$

From *NCEES Handbook:* F-Factor (Log-Mean Temperature Correction Factor or LMTC Factor), the temperature effectiveness, P, and heat capacity rate ratio, R, required by the LMTD correction factor correlation are

$$R = \frac{T_{\text{shell,in}} - T_{\text{shell,out}}}{T_{\text{tube,out}} - T_{\text{tube,in}}} = \frac{60^\circ\text{C} - 60^\circ\text{C}}{30.66^\circ\text{C} - 25^\circ\text{C}} = 0$$

$$P = \frac{T_{\text{tube,out}} - T_{\text{tube,in}}}{T_{\text{shell,in}} - T_{\text{tube,in}}} = \frac{30.66^\circ\text{C} - 25^\circ\text{C}}{60^\circ\text{C} - 25^\circ\text{C}} = 0.1617$$

From a chart of LMTD correction factors, with these values of R and P, the correction factor, F_c, for one pass on the shell and two passes on the tube-side heat exchanger is 1.0. From *NCEES Handbook:* F-Factor (Log-Mean Temperature Correction Factor or LMTC Factor), the F-Factor, the corrected temperature is

$$\begin{aligned} \Delta T_{\text{mean}} &= F\Delta T_{\text{log mean}} \\ &= (1.0)(32.087^\circ\text{C}) \\ &= 32.087^\circ\text{C} \end{aligned}$$

From *NCEES Handbook:* Combination of Heat-Transfer Mechanisms,

$$\dot{Q} = U_{\text{ov}}A\Delta T$$

The total heating surface of the heat exchanger is

$$A = \frac{\dot{Q}}{U_{ov}\Delta T_{\text{log mean}}} = \frac{8.53 \times 10^8 \text{ W}}{\left(1010 \frac{\text{W}}{\text{m}^2\cdot\text{K}}\right)(32.087°\text{C})} = 26\,321 \text{ m}^2$$

The length of the tubes per pass is

$$L = \frac{A}{2\pi DN} = \frac{26\,321 \text{ m}^2}{(2\pi)(0.05 \text{ m})(10\,000)} = 8.38 \text{ m} \quad (8.4 \text{ m})$$

The answer is (C).

74. The relevant data are shown in the illustration.

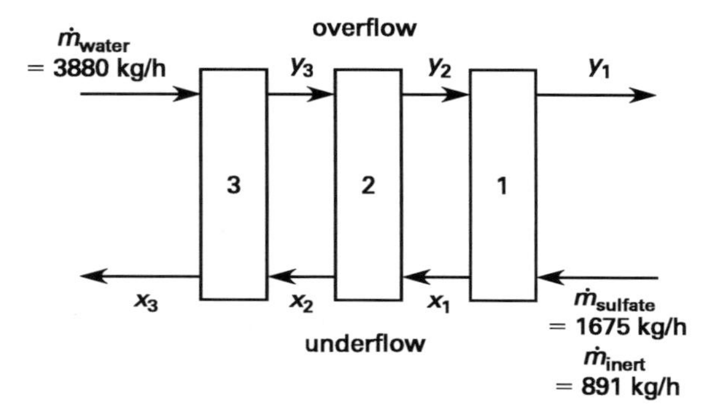

The underflow from each stage is a slurry of inert solids and Na_2SO_4 solution. The mass fraction of Na_2SO_4 in the solution that leaves stage j as part of the slurry is X_j. The mass fraction of Na_2SO_4 solution in the slurry is constant at 0.52. The underflow ratio—the ratio of the mass of solution to the mass of inert solids in the underflow—is

$$R = \frac{m_{\text{solution}}}{m_{\text{inert}}} = \frac{0.52}{1 - 0.52} = 1.083$$

The liquid overflow from each stage consists solely of Na_2SO_4 solution and contains none of the inert material; the mass fraction of Na_2SO_4 in the solution leaving stage j is Y_j.

The liquid overflow contains none of the inert solids, so the mass flow rate of the inert solids is constant throughout the process. The mass fraction of solution in the slurry is also constant, so the total mass flow rate of the slurry is constant from its creation in stage 1. Because the mass entering each stage must be equal to the mass leaving the same stage, the mass flow rate of the liquid that enters and leaves each stage must also be constant (and therefore equal to $\dot{m}_{\text{water}}$) until stage 1, at which point an amount of solution equal to $R\dot{m}_{\text{inert}}$ mixes with the solids to form the slurry. The liquid overflow from stage 1 that leaves the process is therefore equal to $\dot{m}_{\text{water}} - R\dot{m}_{\text{inert}}$.

From *NCEES Handbook:* Multistage Leaching, the leaching calculation are based on

$$\textit{Inputs} = \textit{Outputs}$$

Each stage reaches equilibrium, so the concentration of Na_2SO_4 in the liquid overflow from each stage is equal to the concentration of Na_2SO_4 in the liquid part of the underflow from the same stage. So,

$$X_1 = Y_1$$
$$X_2 = Y_2$$
$$X_3 = Y_3$$

A mass balance of Na_2SO_4 around stage 1 gives

$$\text{sulfate entering stage 1} = \text{sulfate leaving stage 1}$$

$$Y_2\dot{m}_{\text{water}} + \dot{m}_{\text{sulfate}} = Y_1(\dot{m}_{\text{water}} - R\dot{m}_{\text{inert}}) + X_1R\dot{m}_{\text{inert}}$$
$$X_2\dot{m}_{\text{water}} + \dot{m}_{\text{sulfate}} = X_1\dot{m}_{\text{water}} - X_1R\dot{m}_{\text{inert}} + X_1R\dot{m}_{\text{inert}}$$
$$X_1 = X_2 + \frac{\dot{m}_{\text{sulfate}}}{\dot{m}_{\text{water}}} = X_2 + \frac{1675 \frac{\text{kg}}{\text{h}}}{3880 \frac{\text{kg}}{\text{h}}} = X_2 + 0.432$$

A mass balance of Na_2SO_4 around stage 2 gives

$$Y_3\dot{m}_{\text{water}} + X_1R\dot{m}_{\text{inert}} = Y_2\dot{m}_{\text{water}} + X_2R\dot{m}_{\text{inert}}$$
$$X_3\dot{m}_{\text{water}} + (X_2 + 0.432)\times(R\dot{m}_{\text{inert}}) = X_2\dot{m}_{\text{water}} + X_2R\dot{m}_{\text{inert}}$$
$$X_2 = X_3 + \frac{0.432R\dot{m}_{\text{inert}}}{\dot{m}_{\text{water}}} = X_3 + \frac{(0.432)(1.083)\times\left(891 \frac{\text{kg}}{\text{h}}\right)}{3880 \frac{\text{kg}}{\text{h}}} = X_3 + 0.107$$

The liquid entering stage 3 is pure water with no Na_2SO_4, so a mass balance of Na_2SO_4 around stage 3 gives

$$\begin{aligned} X_2 R\dot{m}_{\text{inert}} &= Y_3 \dot{m}_{\text{water}} + X_3 R\dot{m}_{\text{inert}} \\ (X_3 + 0.107) R\dot{m}_{\text{inert}} &= X_3 \dot{m}_{\text{water}} + X_3 R\dot{m}_{\text{inert}} \\ X_3 &= \frac{0.107 R\dot{m}_{\text{inert}}}{\dot{m}_{\text{water}}} \\ &= \frac{(0.107)(1.083)\left(891\ \frac{\text{kg}}{\text{h}}\right)}{3880\ \frac{\text{kg}}{\text{h}}} \\ &= 0.0266 \end{aligned}$$

The mass fraction of Na_2SO_4 in the liquid overflow leaving stage 1 is

$$\begin{aligned} Y_1 = X_1 &= X_2 + 0.432 \\ &= (X_3 + 0.107) + 0.432 \\ &= 0.0266 + 0.107 + 0.432 \\ &= 0.566 \end{aligned}$$

The recovery of Na_2SO_4 is

$$\begin{aligned} \text{recovery} &= \frac{Y_1(\dot{m}_{\text{water}} - R\dot{m}_{\text{inert}})}{\dot{m}_{\text{sulfate}}} \\ &= \frac{(0.566)\left(3880\ \frac{\text{kg}}{\text{h}} - (1.083)\left(891\ \frac{\text{kg}}{\text{h}}\right)\right)}{1675\ \frac{\text{kg}}{\text{h}}} \\ &= 0.985 \quad (0.99) \end{aligned}$$

The answer is (D).

75. Throughout this solution the *NCEES Handbook:* Material Balances balance equation is used extensively.

$$\begin{aligned} \textit{Accumulation} &= \textit{Input} - \textit{Output} \\ &\quad + \textit{Generation} - \textit{Consumption} \end{aligned}$$

For each component in the outlet stream, find the balanced reaction for complete combustion. Base all calculations on 1.0 mol of fuel fed into the combustion chamber. The reaction is

$$CH_4 + 2O_2 \rightarrow CO_2 + 2H_2O$$

Therefore, all the fuel is consumed during the combustion. The mole fraction of methane, CH_4, in the fuel is 0.61, so in each 1.0 mol of fuel, the number of moles of CH_4 that react is

$$n_{CH_4,\text{fuel}} = (1.0\ \text{mol})(0.61) = 0.61\ \text{mol}$$

From the balanced reaction for the complete combustion of CH_4, all the carbon from the CH_4 becomes part of the carbon dioxide, CO_2, so the number of moles of CO_2 produced from the CH_4 in 1.0 kmol of fuel must also be 0.61 mol.

$$n_{CO_2\ \text{from}\ CH_4} = n_{CH_4,\text{fuel}} = 0.61\ \text{mol}$$

All the hydrogen from the CH_4 becomes part of the water, H_2O, so the number of moles of H_2O produced from the CH_4 in 1.0 mol of fuel is twice the number of moles of CH_4, or

$$n_{H_2O\ \text{from}\ CH_4} = 2n_{CH_4,\text{fuel}} = (2)(0.61\ \text{mol}) = 1.22\ \text{mol}$$

The balanced reaction for complete combustion of propane, C_3H_8, is

$$C_3H_8 + 5O_2 \rightarrow 3CO_2 + 4H_2O$$

All the fuel is consumed during the combustion, and the mole fraction of the C_3H_8 in the fuel is 0.36, so in each 1.0 mol of fuel, the number of moles of C_3H_8 that react is

$$n_{C_3H_8,\text{fuel}} = (1.0\ \text{mol})(0.36) = 0.36\ \text{mol}$$

From the balanced reaction for the complete combustion of C_3H_8, all the carbon from the C_3H_8 becomes part of the CO_2, so the number of moles of CO_2 produced from the C_3H_8 in 1.0 mol of fuel must be three times the number of moles of C_3H_8, or

$$n_{CO_2\ \text{from}\ C_3H_8} = 3n_{C_3H_8,\text{fuel}} = (3)(0.36\ \text{mol}) = 1.08\ \text{mol}$$

All the hydrogen from the C_3H_8 becomes part of the H_2O, so the number of moles of H_2O produced from the C_3H_8 in 1.0 mol of fuel is four times the number of moles of C_3H_8, or

$$n_{H_2O\ \text{from}\ C_3H_8} = 4n_{C_3H_8,\text{fuel}} = (4)(0.36\ \text{mol}) = 1.44\ \text{mol}$$

The total CO_2 in the outlet stream produced by 1.0 mol of fuel is

$$\begin{aligned} n_{CO_2,\text{outlet}} &= n_{CO_2\ \text{from}\ CH_4} + n_{CO_2\ \text{from}\ C_3H_8} \\ &= 0.61\ \text{mol} + 1.08\ \text{mol} \\ &= 1.69\ \text{mol} \end{aligned}$$

From the left side of the balanced combustion reaction of CH_4, twice as many moles of oxygen, O_2, are needed as moles of CH_4. The number of moles of O_2 needed for complete combustion of 0.61 mol of CH_4 is

$$n_{O_2\ \text{needed for}\ CH_4} = 2n_{CH_4,\text{fuel}} = (2)(0.61\ \text{mol}) = 1.22\ \text{mol}$$

From the left side of the balanced combustion reaction of C_3H_8, five times as many moles of O_2 are needed as moles of C_3H_8. The number of moles of O_2 needed for complete combustion of 0.36 mol of C_3H_8 is

$$n_{O_2 \text{ needed for } C_3H_8} = 5n_{C_3H_8,\text{fuel}} = (5)(0.36 \text{ mol}) = 1.80 \text{ mol}$$

From *NCEES Handbook:* Combustion Reactions, the theoretical (stoichiometric) is the air required for complete combustion. Therefore, the number of moles of O_2 needed for complete combustion of the CH_4 and C_3H_8 in 1.0 mol of the feed is

$$\begin{aligned} n_{O_2,\text{needed}} &= n_{O_2 \text{ needed for } CH_4} + n_{O_2 \text{ needed for } C_3H_8} \\ &= 1.22 \text{ mol} + 1.80 \text{ mol} \\ &= 3.02 \text{ mol} \end{aligned}$$

25% excess air is used, so

$$\begin{aligned} n_{O_2,\text{excess}} &= 0.25n_{O_2,\text{needed}} = (0.25)(3.02 \text{ mol}) \\ &= 0.755 \text{ mol} \end{aligned}$$

The excess air leaves the combustion chamber unused, so

$$n_{O_2,\text{outlet}} = n_{O_2,\text{excess}} = 0.755 \text{ mol}$$

The number of moles of O_2 actually used is

$$\begin{aligned} n_{O_2,\text{inlet}} &= n_{O_2,\text{needed}} + n_{O_2,\text{excess}} \\ &= 3.02 \text{ mol} + 0.755 \text{ mol} \\ &= 3.775 \text{ mol} \end{aligned}$$

Air is 21% O_2, so the number of moles of air used during the combustion is

$$n_{\text{inlet air}} = \frac{n_{O_2,\text{inlet}}}{0.21} = \frac{3.775 \text{ mol}}{0.21} = 17.98 \text{ mol}$$

From *NCEES Handbook* table "Selected Properties of Air," the molecular weight of air, MW_{air}, is 28.965 lbm/lb mole or 28.965 g/mol. The molecular weight of air, MW_{air}, is given as 29 g/mol. The mass of air used for combustion of 1.0 kmol of the feed is

$$\begin{aligned} m_{\text{inlet air}} &= n_{\text{inlet air}}(MW_{\text{air}}) \\ &= (17.98 \text{ mol})\left(28.965 \ \frac{\text{g}}{\text{mol}}\right) \\ &= 520.79 \text{ g} \end{aligned}$$

The air into the combustion chamber is at a dry-bulb temperature of 36°C and a wet-bulb temperature of 32°C. As in *NCEES Handbook:* Physical Properties of Air, the absolute humidity of the air, ω, is 0.029 g of water/g dry air. The mass of water in the inlet air is

$$\begin{aligned} m_{H_2O,\text{inlet air}} &= \omega m_{\text{inlet air}} \\ &= \left(0.029 \ \frac{\text{g}}{\text{g}}\right)(520.79 \text{ g}) \\ &= 15.10 \text{ g} \end{aligned}$$

The molecular weight of water is given as 18 g/mol. From *NCEES Handbook:* Behavior of Multicomponent Systems, molecular weight is

$$MW = \frac{m}{n}$$

The number of moles of water introduced with the inlet air is

$$n_{H_2O,\text{inlet air}} = \frac{m_{H_2O,\text{inlet air}}}{MW_{H_2O}} = \frac{15.10 \text{ g}}{18 \ \frac{\text{g}}{\text{mol}}} = 0.839 \text{ mol}$$

The water in the outlet stream comes from three sources: the water formed in the combustion of CH_4, the water produced in the combustion of C_3H_8, and the water in the inlet air. The total number of moles of water in the outlet stream is

$$\begin{aligned} n_{H_2O,\text{outlet}} &= n_{H_2O \text{ from } CH_4} + n_{H_2O \text{ from } C_3H_8} + n_{H_2O,\text{inlet air}} \\ &= 1.22 \text{ mol} + 1.44 \text{ mol} + 0.838 \text{ mol} \\ &= 3.498 \text{ mol} \end{aligned}$$

Air is 79% nitrogen, N_2, so the number of moles of N_2 in the inlet air is

$$\begin{aligned} n_{N_2,\text{inlet air}} &= 0.79n_{\text{inlet air}} \\ &= (0.79)(17.98 \text{ mol}) \\ &= 14.20 \text{ mol} \end{aligned}$$

The fuel gas contains only CH_4, C_3H_8, and N_2, so the number of moles of N_2 in 1.0 mol of fuel gas is

$$\begin{aligned} n_{N_2,\text{fuel}} &= n_{\text{fuel}} - n_{CH_4,\text{fuel}} - n_{C_3H_8,\text{fuel}} \\ &= 1.0 \text{ mol} - 0.61 \text{ mol} - 0.36 \text{ mol} \\ &= 0.03 \text{ mol} \end{aligned}$$

The N_2 does not react in the combustion chamber, so the N_2 that enters the chamber leaves it unchanged.

$$\begin{aligned} n_{N_2,\text{outlet}} &= n_{N_2,\text{inlet}} = n_{N_2,\text{inlet air}} + n_{N_2,\text{fuel}} \\ &= 14.20 \text{ mol} + 0.03 \text{ mol} \\ &= 14.23 \text{ mol} \end{aligned}$$

The outlet stream consists of CO_2, H_2O, O_2, and N_2.

The number of moles in the outlet stream per mole of fuel in the feed is

$$\begin{aligned} n_{\text{outlet}} &= n_{CO_2,\text{outlet}} + n_{H_2O,\text{outlet}} + n_{O_2,\text{outlet}} + n_{N_2,\text{outlet}} \\ &= 1.69 \text{ mol} + 3.498 \text{ mol} + 0.755 \text{ mol} + 14.23 \text{ mol} \\ &= 20.173 \text{ mol} \quad (20 \text{ mol}) \end{aligned}$$

The answer is (D).

76. From *NCEES Handbook:* Net-Positive Suction Head (NPSH), the formula for net-positive suction head available to the pump is

$$NPSH_a = h_a - h_{\text{vap}} - h_{\text{st}} - h_L$$

The liquid level in the tank is 10 ft higher than the pump. Define z_{pump} as 0 ft and z_{surface} as 10 ft. The static head, h_s, is the sum of the atmospheric pressure head, h_p, and the static suction head, $h_{z(s)}$.

$$\begin{aligned} h_s &= h_p + h_{z(s)} \\ &= h_p + (z_{\text{surface}} - z_{\text{pump}}) \\ &= h_p + 10 \text{ ft} \end{aligned}$$

The density of the fluid, ρ, is given as 60.58 lbm/ft^3. As the tank is vented to the atmosphere, the pressure, P, is 14.7 lbf/in^2. The atmospheric pressure head is

$$\begin{aligned} h_p &= \frac{Pg_c}{\rho g} \\ &= \frac{\left(14.7 \frac{\text{lbf}}{\text{in}^2}\right)\left(12 \frac{\text{in}}{\text{ft}}\right)^2\left(32.2 \frac{\text{ft-lbm}}{\text{lbf-sec}^2}\right)}{\left(60.58 \frac{\text{lbm}}{\text{ft}^3}\right)\left(32.2 \frac{\text{ft}}{\text{sec}^2}\right)} \\ &= 34.94 \text{ ft} \end{aligned}$$

The static head is

$$h_s = h_p + 10 \text{ ft} = 34.94 \text{ ft} + 10 \text{ ft} = 44.94 \text{ ft}$$

The vapor pressure head, h_{vap}, is found from the vapor pressure and density of the fluid. The vapor pressure of the fluid, P_v, is given as 7.51 lbf/in^2. The vapor pressure head is

$$\begin{aligned} h_{\text{vap}} &= \frac{P_v g_c}{\rho g} \\ &= \frac{\left(7.51 \frac{\text{lbf}}{\text{in}^2}\right)\left(12 \frac{\text{in}}{\text{ft}}\right)^2\left(32.2 \frac{\text{ft-lbm}}{\text{lbf-sec}^2}\right)}{\left(60.58 \frac{\text{lbm}}{\text{ft}^3}\right)\left(32.2 \frac{\text{ft}}{\text{sec}^2}\right)} \\ &= 17.85 \text{ ft} \end{aligned}$$

The equivalent length of the suction line, L, is 52.26 ft. The velocity of the fluid at the pump is found from the formula $u = \dot{Q}/A$. The inside diameter of 4 in schedule-40 pipe is

$$D = \frac{4.026 \text{ in}}{12 \frac{\text{in}}{\text{ft}}} = 0.3355 \text{ ft}$$

The cross-sectional area of the pipe is

$$\begin{aligned} A &= \frac{\pi D^2}{4} \\ &= \frac{\pi (0.3355 \text{ ft})^2}{4} \\ &= 0.0884 \text{ ft}^2 \end{aligned}$$

The flow rate, $\dot{Q}$, is 180 gal/min, so the velocity of the water at the pump is

$$\begin{aligned} u &= \frac{\dot{Q}}{A} \\ &= \frac{180 \frac{\text{gal}}{\text{min}}}{(0.0884 \text{ ft}^2)\left(7.48 \frac{\text{gal}}{\text{ft}^3}\right)\left(60 \frac{\text{sec}}{\text{min}}\right)} \\ &= 4.537 \text{ ft/sec} \end{aligned}$$

To find the friction factor, the Reynolds number and relative roughness are needed. The viscosity of the fluid is

$$\begin{aligned} \mu &= (0.35 \text{ cP})\left(6.72 \times 10^{-4} \frac{\frac{\text{lbm}}{\text{ft-sec}}}{\text{cP}}\right) \\ &= 0.000235 \text{ lbm/ft-sec} \end{aligned}$$

From *NCEES Handbook* table "Dimensionless Numbers," the Reynolds number is

$$\begin{aligned} Re &= \frac{\rho u D}{\mu} \\ &= \frac{\left(60.58 \frac{\text{lbm}}{\text{ft}^3}\right)\left(4.537 \frac{\text{ft}}{\text{sec}}\right)(0.3355 \text{ ft})}{0.000235 \frac{\text{lbm}}{\text{ft-sec}}} \\ &= 392{,}394.3 \end{aligned}$$

From *NCEES Handbook:* Absolute Roughness and Relative Roughness, the roughness of the pipe, ε, is given as 0.00015 ft. The relative roughness of the pipe is

$$\frac{\varepsilon}{D} = \frac{0.00015 \text{ ft}}{0.3355 \text{ ft}} = 0.000447$$

Applying the Reynolds number and relative roughness to the Moody diagram, the friction factor is found to be 0.0176. From *NCEES Handbook:* Head Loss in Pipe or Conduit, the friction head loss, h_f, is found by means of the Darcy-Weisbach equation.

$$\begin{aligned} h_L &= f\frac{L}{D}\frac{u^2}{2g} \\ &= (0.0176)\frac{(52.26 \text{ ft})\left(4.537 \ \frac{\text{ft}}{\text{sec}}\right)^2}{(0.3355 \text{ ft})(2)\left(32.2 \ \frac{\text{ft}}{\text{sec}^2}\right)} \\ &= 0.8763 \text{ ft} \end{aligned}$$

The head loss due to a sudden contraction where the fluid leaves the tank, h_c, is found from the resistance coefficient for the sudden contraction, K, which is given as 0.4. From *NCEES Handbook:* Head Loss in Pipe or Conduit, the head loss due to a sudden contraction is

$$\begin{aligned} h_c &= \frac{Ku^2}{2g} = \frac{(0.4)\left(4.537 \ \frac{\text{ft}}{\text{sec}}\right)^2}{(2)\left(32.2 \ \frac{\text{ft}}{\text{sec}^2}\right)} \\ &= 0.128 \text{ ft} \end{aligned}$$

The net positive suction head is therefore

$$\begin{aligned} NPSH_a &= h_s - h_{vap} - h_L - h_c \\ &= 44.94 \text{ ft} - 17.85 \text{ ft} - 0.8763 \text{ ft} - 0.128 \text{ ft} \\ &= 26.086 \text{ ft} \quad (26 \text{ ft}) \end{aligned}$$

The answer is (B).

77. From *NCEES Handbook* table "Physical Constants," the acceleration due to gravity, g, is 9.8067 m/s^2. The absolute temperature of the air surrounding the pipe is

$$T_{air} = 20°\text{C} + 273° = 293\text{K}$$

The absolute temperature of the surface of the pipe is

$$T_{surface} = 234°\text{C} + 273° = 507\text{K}$$

The internal diameter of the pipe, D, is 0.1 m. From *NCEES Handbook* table "Dimensionless Numbers," the Rayleigh number is

$$\begin{aligned} Ra &= \frac{g\beta\Delta TL^3}{\nu^2}Pr = \frac{g\beta(T_{surface} - T_{air})D^3}{\nu\alpha} \\ &= \frac{\left(9.8067 \ \frac{\text{m}}{\text{s}^2}\right)(2.73 \times 10^{-3} \text{ K}^{-1})(507\text{K} - 293\text{K})(0.1 \text{ m})^3}{\left(2.64 \times 10^{-5} \ \frac{\text{m}^2}{\text{s}}\right)\left(3.83 \times 10^{-5} \ \frac{\text{m}^2}{\text{s}}\right)} \\ &= 5.66 \times 10^6 \end{aligned}$$

The average Nusselt number is

$$\begin{aligned} \overline{\text{Nu}} &= \left(0.60 + \frac{0.387(\text{Ra})^{1/6}}{\left(1 + \left(\frac{0.559}{Pr}\right)^{9/16}\right)^{8/27}}\right)^2 \\ &= \left(0.60 + \frac{(0.387)(5.66 \times 10^6)^{1/6}}{\left(1 + \left(\frac{0.559}{0.697}\right)^{9/16}\right)^{8/27}}\right)^2 \\ &= 23.84 \end{aligned}$$

From *NCEES Handbook* table "Forced Convection—External Flow," the average Nusselt number is

$$\overline{\text{Nu}} = \frac{\bar{h}D}{k}$$

Solving the preceding equation for the convection heat transfer coefficient

$$\begin{aligned} \bar{h} &= \left(\frac{k}{D}\right)\overline{\text{Nu}} \\ &= \left(\frac{0.0338 \ \frac{\text{W}}{\text{m·K}}}{0.1 \text{ m}}\right)(23.84) \\ &= 8.058 \text{ W/m}^2\text{·K} \end{aligned}$$

The heat loss due to convection is

$$\begin{aligned} q_{conv} &= \bar{h}\pi D(T_{surface} - T_{air}) \\ &= \left(8.058 \ \frac{\text{W}}{\text{m}^2\text{·K}}\right)\pi(0.1 \text{ m})(507\text{K} - 293\text{K}) \\ &= 541.74 \text{ W/m} \end{aligned}$$

As in *NCEES Handbook:* Stefan-Boltzmann Law of Radiation, the heat loss due to radiation is

$$\begin{aligned} q_{\text{rad}} &= \varepsilon \pi D \sigma (T_{\text{surface}}^4 - T_{\text{air}}^4) \\ &= (0.80)\pi(0.1 \text{ m})\left(5.67 \times 10^{-8} \frac{\text{W}}{\text{m}^2{\cdot}\text{K}^4}\right) \\ &\quad \times \left((507\text{K})^4 - (293\text{K})^4\right) \\ &= 836.55 \text{ W/m} \end{aligned}$$

σ is the Stefan-Boltzmann constant. The total heat loss is the sum of the heat losses due to convection and radiation.

$$\begin{aligned} q_{\text{total}} &= q_{\text{conv}} + q_{\text{rad}} \\ &= 541.74 \frac{\text{W}}{\text{m}} + 836.55 \frac{\text{W}}{\text{m}} \\ &= 1378.29 \text{ W/m} \quad (1400 \text{ W/m}) \end{aligned}$$

The answer is (D).

78. The relevant data are shown in the illustration.

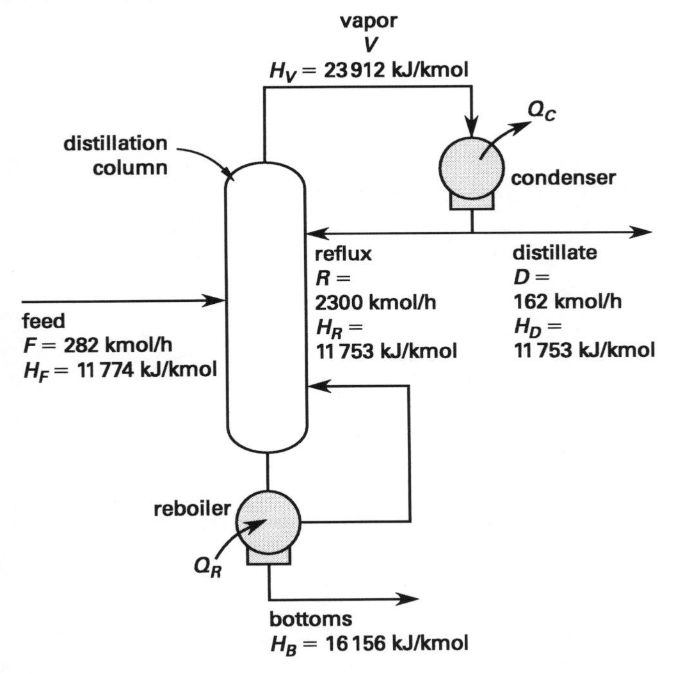

As in *NCEES Handbook:* Material and Energy Balances for Trayed and Packed Units, a mass balance around the splitter gives the molar flow rate of the overhead stream.

$$\begin{aligned} \dot{V} &= \dot{R} + \dot{D} \\ &= 2300 \frac{\text{kmol}}{\text{h}} + 162 \frac{\text{kmol}}{\text{h}} \\ &= 2462 \text{ kmol/h} \end{aligned}$$

From *NCEES Handbook:* Column Material Balance, a mass balance around the entire system gives the molar flow rate of the bottoms stream.

$$F = D + B$$

Or

$$\begin{aligned} \dot{B} &= \dot{F} - \dot{D} = 282 \frac{\text{kmol}}{\text{h}} - 162 \frac{\text{kmol}}{\text{h}} \\ &= 120 \text{ kmol/h} \end{aligned}$$

As in *NCEES Handbook:* Condensers, an energy balance around the condenser gives the condenser duty.

$$\begin{aligned} \dot{Q}_C &= \dot{V}(\hat{h}_V - \hat{h}_R) \\ &= \left(2462 \frac{\text{kmol}}{\text{h}}\right)\left(23\,912 \frac{\text{kJ}}{\text{kmol}} - 11\,753 \frac{\text{kJ}}{\text{kmol}}\right) \\ &= 29\,935\,458 \text{ kJ/h} \end{aligned}$$

As in *NCEES Handbook:* Material and Energy Balances for Trayed and Packed Units, an energy balance around the distillation column gives the reboiler duty.

$$\begin{aligned} \dot{Q}_R &= \dot{D}\hat{h}_D + \dot{B}\hat{h}_B + \dot{Q}_C - \dot{F}\hat{h}_F \\ &= \left(162 \frac{\text{kmol}}{\text{h}}\right)\left(11\,753 \frac{\text{kJ}}{\text{kmol}}\right) + \left(120 \frac{\text{kmol}}{\text{h}}\right) \\ &\quad \times \left(16\,156 \frac{\text{kJ}}{\text{kmol}}\right) + 29\,935\,458 \frac{\text{kJ}}{\text{h}} \\ &\quad - \left(282 \frac{\text{kmol}}{\text{h}}\right)\left(11\,774 \frac{\text{kJ}}{\text{kmol}}\right) \\ &= 30\,457\,896 \text{ kJ/h} \quad (3.0 \times 10^7 \text{ kJ/h}) \end{aligned}$$

The answer is (C).

79. The charge, n_0, is 100 kmol. The total pressure, converted to millimeters of mercury, is

$$\begin{aligned} P &= (1.2 \text{ atm})\left(760 \frac{\text{mm Hg}}{\text{atm}}\right) \\ &= 912 \text{ mm Hg} \end{aligned}$$

As in *NCEES Handbook:* Distribution of Components Between Phases in a Vapor/Liquid Equilibrium, the distribution coefficient for styrene is

$$\begin{aligned} K_s &= \frac{P_s}{P} \\ &= \frac{e^{14.3284 - \left(3516.43\text{K}/(T - 56.1529\text{K})\right)}}{912 \text{ mm Hg}} \end{aligned}$$

Similarly, the distribution coefficient for toluene is

$$K_t = \frac{P_t}{P} = \frac{e^{14.2515-\left(3242.38\text{K}/(T-47.1806\text{K})\right)}}{912 \text{ mm Hg}}$$

From *NCEES Handbook:* Bubble Point, the bubble point in terms of the vapor-liquid equilibrium is defined by

$$\sum_{i=1}^{n} K_i x_i = 1$$

$$x_s K_s + x_t K_t = 1.0$$

Replacing,

$$(0.62)\left(\frac{e^{14.3284-\left(3516.43\text{K}/(T-56.1529\text{K})\right)}}{912 \text{ mm Hg}}\right) + (0.38)\left(\frac{e^{14.2515-\left(3242.38\text{K}/(T-47.1806\text{K})\right)}}{912 \text{ mm Hg}}\right) = 1.0$$

Solving for the bubble point (using a programmable calculator) gives

$$T = 505.571\text{K}$$

Substituting this value for T in the equation for K_s gives

$$\begin{aligned} K_s &= \frac{e^{14.3284-\left(3516.43\text{K}/(T-56.1529\text{K})\right)}}{912 \text{ mm Hg}} \\ &= \frac{e^{14.3284-\left(3516.43\text{K}/(505.571\text{K}-56.1529\text{K})\right)}}{912 \text{ mm Hg}} \\ &= 0.73224 \end{aligned}$$

The desired composition at the end of the distillation is 15% toluene, so the mole fraction of toluene in the liquid remaining in the still, x_{t2}, is 0.15. The liquid remaining in the still consists of two components, styrene and toluene. The mole fraction of styrene in the liquid remaining in the still is

$$\begin{aligned} x_{s2} &= 1 - x_{t2} \\ &= 1 - 0.15 \\ &= 0.85 \end{aligned}$$

The vapor composition is in equilibrium with the perfectly mixed liquid in the still.

$$y_s^* = K_s x_s$$

The Rayleigh equation applies for the amount of liquid remaining in the still, n. For styrene, from *NCEES Handbook:* Rayleigh Equation, the Rayleigh equation is

$$\int_{n_0}^{n_\text{f}} \frac{dn}{n} = \ln\frac{n_\text{f}}{n_0} = \int_{x_0}^{x_\text{f}} \frac{dx}{y - x}$$

$$\frac{dn}{dx_s} = \frac{n}{y_s^* - x_s}$$

Substituting the equation for y_s^* into the Rayleigh equation gives

$$\begin{aligned} \frac{dn}{dx_s} &= \frac{n}{K_s x_s - x_s} = \frac{n}{x_s(k_s - 1)} \\ \frac{dn}{n} &= \frac{dx_s}{x_s(k_s - 1)} \\ &= \frac{dx_s}{x_s(0.73223 - 1)} \\ &= \frac{dx_s}{x_s(-0.26777)} \end{aligned}$$

Integrating,

$$\begin{aligned} \int_{n_0}^{n_\text{f}} \frac{dn}{n} &= \int_{x_s}^{x_{s2}} \frac{dx_s}{x_s(-0.26777)} = \int_{0.62}^{0.85} \frac{dx_s}{x_s(-0.26777)} \\ &= \frac{\ln\frac{0.85}{0.62}}{-0.26777} \\ &= -1.17831 \\ \ln\frac{n_\text{f}}{n_0} &= -1.17831 \\ n_\text{f} &= e^{-1.17831+\ln n_0} \\ &= e^{-1.17831+\ln 100} \\ &= 30.78 \text{ kmol} \quad (31 \text{ kmol}) \end{aligned}$$

The answer is (D).

80. Options A, B, D, and F all meet one or more of the characteristics of a rupture disk installed in series with a relief valve. Options C and E do not meet these conditions.

The answers are A, B, D, and F.